Progress in Inflammation Research

Forthcoming titles:

New Therapeutic Targets in Rheumatoid Arthritis

Paul-Peter Tak

Editor

Birkhäuser
Basel · Boston · Berlin

Editor

Paul-Peter Tak
Division of Clinical Immunology and Rheumatology
EULAR & FOCIS Center of Excellence
Academic Medical Center
University of Amsterdam
Meibergdreef 9, Room F4-218
1105 AZ Amsterdam
The Netherlands

Library of Congress Control Number: 2009921933

Bibliographic information published by Die Deutsche Bibliothek
Die Deutsche Bibliothek lists this publication in the Deutsche Nationalbibliografie;
detailed bibliographic data is available in the internet at http://dnb.ddb.de

ISBN 978-3-7643-8237-7 Birkhäuser Verlag AG, Basel – Boston – Berlin

© 2009 Birkhäuser Verlag AG
Basel · Boston · Berlin
P.O. Box 133, CH-4010 Basel, Switzerland
Part of Springer Science+Business Media
Printed on acid-free paper produced from chlorine-free pulp. TCF ∞
Cover design: Markus Etterich, Basel
Cover illustration: see page 68. With friendly permission of Wim B. van den Berg.
Printed in Germany
ISBN 978-3-7643-8237-7   e-ISBN 978-3-7643-8238-4
9 8 7 6 5 4 3 2 1   www.birkhauser.ch

Contents

List of contributors

Theresa C. Barnes, Division of Inflammation, School of Clinical Sciences, University of Liverpool, University Hospital Aintree, Longmoor Lane, Liverpool L9 7AL, UK

Shouvik Dass, Section of Musculoskeletal Disease, Leeds Institute of Molecular Medicine, University of Leeds, Chapel Allerton Hospital, Leeds LS7 4SA, UK

Charles A. Dinarello, Department of Medicine, Division of Infectious Diseases, University of Colorado Denver, 12700 East 19th Ave, B168, Aurora, CO 80045, USA; e-mail: cdinare333@aol.com

Paul Emery, Section of Musculoskeletal Disease, Leeds Institute of Molecular Medicine, University of Leeds, Chapel Allerton Hospital, Leeds LS7 4SA, UK; e-mail: p.emery@leeds.ac.uk

Gary S. Firestein, University of California, San Diego, Department of Medicine, Biomedical Science Bldg. RM 5096, 9500 Gilman Drive, La Jolla, CA 92093-0656, USA; e-mail: gfirestein@ucsd.edu

Steffen Gay, Center of Experimental Rheumatology, Department of Rheumatology, University Hospital of Zürich, Zürich Center of Integrative Human Physiology (ZIHP), Gloriastrasse 25, 8091 Zürich, Switzerland

Lars C. Huber, Center of Experimental Rheumatology, Department of Rheumatology, University Hospital of Zürich, Zürich Center of Integrative Human Physiology (ZIHP), Gloriastrasse 25, 8091 Zürich, Switzerland; e-mail: Lars.Huber@usz.ch

Leo A.B. Joosten, Rheumatology Research & Advanced Therapeutics, Department of Rheumatology, Radboud University Nijmegen Medical Centre, Geert Grooteplein 28, 6525 GA, Nijmegen, The Netherlands

Astrid Jüngel, Center of Experimental Rheumatology, Department of Rheumatology, University Hospital of Zürich, Zürich Center of Integrative Human Physiology (ZIHP), Gloriastrasse 25, 8091 Zürich, Switzerland

Edward C. Keystone, University of Toronto, Division of Rheumatology, Rebecca MacDonald Centre for Arthritis and Autoimmune Diseases, 60 Murray Street, Toronto, Ontario M5T 3L9, Canada; e-mail: edkeystone@mtsinai.on.ca

Tadamitsu Kishimoto, Graduate School of Frontier Bioscience, Osaka University, Osaka 565-0871, Japan; e-mail: kishimot@imed3.med.osaka-u.ac.jp

Alisa E. Koch, Veterans' Administration, Ann Arbor Healthcare System, Ann Arbor, MI, USA and University of Michigan Health System, Department of Internal Medicine, Division of Rheumatology, Ann Arbor, MI, USA

Iain B. McInnes, Centre for Rheumatic Diseases, Division of Immunology, Infection and Inflammation, University of Glasgow, 10 Alexandra Parade, Glasgow Royal Infirmary, Glasgow G312ER, UK; e-mail: i.b.mcinnes@clinmed.gla.ac.uk

Pierre Miossec, Department of Immunology and Rheumatology, Hôpital Edouard Herriot, 69437 Lyon Cedex 03, France; e-mail: miossec@univ-lyon1.fr

Robert J. Moots, Division of Inflammation, School of Clinical Sciences, University of Liverpool, University Hospital Aintree, Longmoor Lane, Liverpool L9 7AL, UK; e-mail: rjmoots@liv.ac.uk

Yoshiyuki Ohsugi, Chugai Pharmaceutical Co. Ltd., Tokyo 103-8324, Japan; e-mail: ohsugiysy@chugai-pharm.co.jp

Richard M. Pope, Division of Rheumatology, Northwestern University Feinberg School of Medicine, McGaw M300, 240 E. Huron Street, Chicago, IL 60611, USA

Eric M. Ruderman, Division of Rheumatology, Northwestern University Feinberg School of Medicine, McGaw M300, 240 E. Huron Street, Chicago, IL 60611, USA

Jagtar Nijar Singh, Centre for Rheumatic Diseases, Division of Immunology, Infection and Inflammation, University of Glasgow, 10 Alexandra Parade, Glasgow Royal Infirmary, Glasgow G312ER, UK

Zoltán Szekanecz, Department of Rheumatology, Institute of Medicine, University of Debrecen Medical and Health Science Center, 22 Móricz street, Debrecen, H-4004, Hungary; e-mail: szekanecz@iiibel.dote.hu

Ling Toh, Department of Immunology and Rheumatology, Hôpital Edouard Herriot, 69437 Lyon Cedex 03, France

Fons A.J. van de Loo, Rheumatology Research & Advanced Therapeutics, Department of Rheumatology, Radboud University Nijmegen Medical Centre, Geert Grooteplein 28, 6525 GA, Nijmegen, The Netherlands

Wim B. van den Berg, Rheumatology Research & Advanced Therapeutics, Department of Rheumatology, Radboud University Nijmegen Medical Centre, Geert Grooteplein 28, 6525 GA, Nijmegen, The Netherlands;
e-mail: w.vandenberg@reuma.umcn.nl

Edward M. Vital, Section of Musculoskeletal Disease, Leeds Institute of Molecular Medicine, University of Leeds, Chapel Allerton Hospital, Leeds LS7 4SA, UK

Jean-Marc Waldburger, University of Geneva, Centre médical universitaire, 1 rue Michel Servet, 1211 Geneva, Switzerland; e-mail: jean-marc.waldburger@unige.ch

Saloua Zrioual, Department of Immunology and Rheumatology, Hôpital Edouard Herriot, 69437 Lyon Cedex 03, France

Preface

During the past decades important breakthroughs have been made in the treatment of rheumatoid arthritis (RA). First, the implementation of low-dose methotrexate and other conventional disease-modifying anti-rheumatic drugs was introduced as an effective treatment. Second, it was recognized that early immunomodulatory treatment is crucial for controlling the disease and its long-term destructive effects more effectively. Parallel advances in research on the pathogenesis of RA and cytokine biology converged in identifying tumor necrosis factor (TNF) as a key factor in inflammation and matrix destruction. The concept arose that elevated TNF concentrations at the sites of inflammation were driving disease pathology, and the removal of excess TNF from sites of inflammation became a therapeutic goal. Clearly, TNF blockade has revolutionized the treatment of RA, as well as other immune-mediated inflammatory diseases. Anti-TNF treatment results in clinical benefit in a significant proportion of the patients, and it has provided proof of concept for the principle of targeted therapy.

Despite the impressive disease-modifying effects of the TNF blockers, not all patients respond, and patients who exhibited an initial response may lose response due to the development of anti-drug antibodies (human anti-chimeric antibodies and human anti-human antibodies, respectively) and perhaps as a result of escape mechanisms related to the disease process. In fact, the majority of the patients still have disease activity in at least one or two actively inflamed joints. In addition, there have been numerous reports of moderate to severe adverse events associated with their continuous use. There is also still some uncertainty as to how long the available anti-rheumatic biologicals can be continuously employed as RA therapies. Thus, there is still a huge unmet need in the current management of RA.

New therapeutic options would include new targeted therapies, perhaps combinations of biological therapies, targeting of synovial fibroblasts, individualized therapies determined by personal profiles of clinical features and biomarkers, and local treatment for persistent joint disease.

The aim of this book is to highlight advances regarding new therapeutic targets in RA. Obviously, not all therapeutic targets that are currently under investigation can be discussed in one book. Therefore, a choice has been made that will allow insight into targeted therapies with novel mechanisms of action that have recently entered the market. In addition, we present examples of new therapeutic strategies that are in preclinical or clinical development. The first three chapters are devoted to new treatments interfering with B cells, T lymphocyte costimulatory pathways and the interleukin (IL)-6 receptor. These chapters summarize the clinical effects and mechanism of action of new anti-rheumatic treatments that have been shown to be effective (rituximab, abatacept, and tocilizumab) and discuss new related approaches. Rituximab, abatacept, and tocilizumab have become useful additions to the available treatments for RA and, together with the TNF blockers, they have also paved the way for the development of new targeted therapeutic strategies. These biological treatments have raised the bar considerably for the efficacy of new treatments, but the fact that roughly one third of the patients does not respond to currently available treatments leaves a major challenge in RA research.

The next six chapters review data on new approaches that have, in part, been tested in patients with RA, but where there is still uncertainty about the clinical effects, or where the results of clinical trials have not yet been published: e.g., targeting IL-1, IL-15, IL-17, IL-18, chemokines, and signaling pathways. Chapter 10 describes oncostatin M, a pleiotropic cytokine with potential utility as a treatment for inflammatory arthritis; clinical trials aimed at blocking this cytokine may be expected in the near future. Chapter 11 is dedicated to a fundamentally different approach: targeting the epigenetic modifications of synovial cells, which is still in the preclinical phase. Finally, chapter 12 provides a more general perspective of the lessons learned from the use of targeted therapies with regard to the utility of animal models of RA, clinical trial design, pharmacodynamics, immunobiology and key pathogenic elements of disease.

I am grateful to all contributors to this volume for sharing their expertise from basic science to clinical trials and *vice versa*; all of them are key opinion leaders who work at the cutting edge of the development of innovative therapies for RA. I would also like to thank Hans Detlef Klüber of Birkhäuser Verlag AG for his patience and support. I hope all readers will be as thrilled as I am with the exciting developments in this field.

Amsterdam, October 2008   Paul P. Tak

B cell targets in rheumatoid arthritis

Edward M. Vital, Shouvik Dass and Paul Emery

Section of Musculoskeletal Disease, Leeds Institute of Molecular Medicine, University of Leeds, Chapel Allerton Hospital, Leeds LS7 4SA, UK

Abstract

B cells are critical to the pathogenesis of rheumatoid arthritis (RA). There is substantial evidence of the efficacy of depletion of B cells in many patients with RA using the first licensed agent, rituximab. Recent research has focused on enhancing efficacy using other targets to inhibit B cell function, including other B cell-depleting antibodies and cytokines critical to B cell function. The rationale for new B cell targets is discussed, as well as clinical data.

Introduction

B cells are central to concepts of rheumatoid arthritis (RA) pathogenesis as well as novel therapeutic approaches. The successful use of B cell-depleting therapies has led to significant insights but many questions remain unanswered. This chapter reviews the scientific concepts underlying the proposed roles of B cells in RA, related factors that influence B cell behaviour and RA disease processes, and considers therapies directed at these targets, both in terms of scientific rationale and clinical results.

B cells develop from haematopoietic stem cells in the bone marrow (BM). They then migrate through the blood to perifollicular, germinal centre and memory compartments in lymphoid tissue, while developing through maturation and activation stages. Mature naïve B cells exit the BM to enter blood and express CD20 while in the circulation; they can also be identified on the basis of their lack of expression for CD27 and variable expression of CD38 (from low to none). Following migration to lymph nodes and germinal centres, naïve B cells become activated and develop a memory phenotype, now expressing CD27+ but not CD38. After further differentiation, memory cells become plasmablasts and return to the BM to become plasma cells. Their phenotype is now CD27++ and CD38++. The fate of plasma cells is a matter of some debate (Fig. 1). The life span of plasma cell varies from a few days to several months, and plasma cells have been divided in two categories, short lived and long lived [1]. The switch from antibody binding to secretion of antibody marks the final transition from B cells to plasmablasts and plasma cells.

New Therapeutic Targets in Rheumatoid Arthritis, edited by Paul-Peter Tak
© 2009 Birkhäuser Verlag Basel/Switzerland

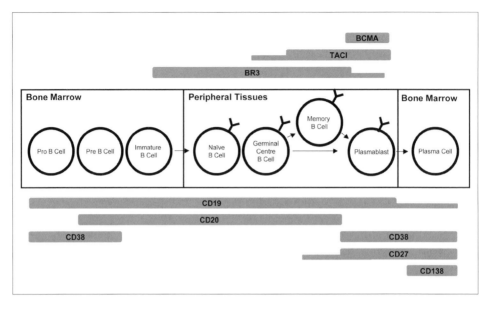

Figure 1
Expression of surface markers and BLyS/APRIL receptors during B cell development.

Antibody secreted by plasma cells has a number of functions. It may directly attack (or stimulate) an antigen; alternatively, binding to antigen may draw in effector mechanisms of the innate immune system such as opsonisation, complement activation and antibody-dependent cellular cytotoxicity (ADCC). However, it is the production of autoantibody by B cells that is thought to lie at the heart of some autoimmune diseases, including RA. Therapies have so far focused on B cell depletion; but while effective in many patients, clinical responses are variable. It is also clear that changes in autoantibody levels do not necessarily correlate with response and relapse. It remains unclear whether better efficacy may be achieved by enhancing depletion further or whether the pattern of eventual B cell repletion is key or, indeed, whether these are linked. Newer therapies in development are attempting to modulate B cell activity rather than simply depleting B cells. These issues are discussed below.

The role of B cells in RA

Current hypotheses of pathogenesis of RA focus on autoantibody production, immune complex generation and the roles of B cells in inflamed synovium. These are discussed below.

Immune complexes in RA

The role of immune complexes as inducers of inflammation in RA was inferred based on their presence in the joint and the consumption of complement specifically in the joints [2, 3], as well as evidence from animal models of inflammatory arthritis. More recently, the distribution of macrophage FcγRIIIa in normal human tissues was found to closely correlate with the pattern of tissue involvement in RA [4]. Binding of this receptor to immune complex gives rise to inflammatory mediators including TNF-α and IL-1α [5]. Animal models of immune complex-mediated arthritis, such as the anti-glucose-6-phosphate isomerase antibodies of the K/BxN mouse, are dependent on the FcγRIII and complement cascade system [6]. Complement component production and activation have been demonstrated in the synovium of rheumatoid, but not osteoarthritis patients [7].

The role of rheumatoid factor

Rheumatoid factor (RF) is present in more than 80% of patients with RA, and may be detectable in the synovium (but not the blood) of some "seronegative RA" patients [8]. Low-affinity "physiological" RF is produced transiently in infectious disease, in which it may facilitate clearance of immune complexes by causing aggregation into larger complexes [9, 10]. In addition to this effect, RF-producing B cells may bind immune complexes containing foreign antigen and present it efficiently to T cells, thus receiving T cell help without needing the presence of an autoreactive, Fc-specific T cell [11]. High-affinity RF-producing B cells that occur in RA may survive and proliferate by a similar mechanism [12].

An early observation in RA was the correlation of the titre of IgM RF with synovial complement consumption [2, 3]. Although IgM RF is most frequent, class-switched RFs are also found, especially in patients with higher RF titres. IgG RFs are able to self-associate, particularly in the joints where there is a relative paucity of normal immunoglobulin (Ig) but large numbers of RF-producing plasma cells [13]. Self-associated IgG RF has also been hypothesized to contribute to pathogenicity of immune complexes in RA by forming smaller, dimeric complexes that evade clearance by complement in blood and access the joint more easily [14].

B cells in the RA synovium

The rheumatoid synovium frequently contains infiltrates of B and T lymphocytes. Their presence, number and histological appearance is, however, variable between patients, and occurs in other diseases besides RA, so the role of these cells within inflamed synovial tissue is complex.

Lymphocytes in the RA synovium are usually arranged in one of three patterns. The majority of patients have a diffuse lymphocytic infiltrate, but substantial numbers have aggregates of B and T cells (without follicular dendritic cells) or germinal centre-like clusters of B and T cells with interdigitating dendritic cells [15–18]. Synovium displaying germinal centre architecture is also characterised by high levels of the cytokines lymphotoxin-α and lymphotoxin-β and the chemokines CCL21 and CXCL13. Lymphoid clusters may be surrounded by plasma cells [15, 19] and the presence of germinal centre-like architecture has been associated with increased synovial production of IgG mRNA [20]. Synovial germinal centre-like structures, however, are not exclusive to RA, and are also observed in diseases not thought to be B cell mediated, such as ankylosing spondylitis [21] and psoriatic arthritis [22]. The presence of larger numbers of plasma cells does, however, appear to be more specific for RA [23].

Synovial B cells often have a mutation pattern of antibody genes characteristic of selection based on antigen specificity and *in situ* B cell differentiation [24, 25]. In addition, sharing of T cell receptor sequences between germinal centres within individual patients is seen [26]. Plasma cells may, therefore, accumulate due to a local germinal centre reaction; however, in early RA they may also be migrating towards chemokines such as SDF-1 [27] and CXCL9 [28]. This latter explanation is also supported by reports of plasma cell infiltration in early arthritis with only small numbers of synovial B cells and no lymphoid follicles [29].

Although anti-CD20 monoclonal antibodies (mAbs) such as rituximab usually deplete peripheral blood B cells profoundly, their effect on synovial B cells is far less consistent, with many patients showing only partial depletion in synovial B cells [30–32]. The synovium therefore appears to be an important niche through which B cells may evade depletion. This may relate to dose and tissue penetration of rituximab, but may also be a function of the microenvironment of the synovium. The factors SDF-1 [33] and BLyS [34] are survival factors for B cells produced in the RA synovium that may protect them from rituximab-mediated apoptosis, as was found in lymphoid tissue in an animal model [35]. Effectiveness of anti-CD20 mAbs against pathogenic plasma cells will depend on targeting these cells' parent lymphocytes and/or cytokines and chemokines involved in plasma cell generation, survival and homing to the joint [36]. Although greater synovial depletion of B cells might be associated with better clinical responses [30], it is not yet clear whether this is related to a downstream effect of *in situ* B cell function and short-lived plasma cell generation or whether it merely reflects reduced B cell trafficking into the joint as a result of reduction of inflammatory cytokines and chemokines, as occurs in other non-B cell-specific biological therapies.

CD20 as a therapeutic target

CD20 is a 33–37-kDa, non-glycosylated phosphoprotein that is expressed on the surface of almost all mature B cells. It has been successfully targeted using the mAb rituximab in a variety of haematological malignancies and rheumatic autoimmune diseases. However, the biology of CD20 and the mechanisms of targeting it are not completely understood. This is in part because CD20 appears to have no natural ligand and also because CD20 knockout mice display an almost normal phenotype [37, 38].

The gene for mouse and human CD20 was completely cloned in 1988 [39–42]. It is located at chromosome 11q12-q13 and consists of eight exons and is 16 kbp long. It is regarded as part of the MS4A gene family [43, 44]. CD20 spans the membrane four times and has two extracellular loops, with the larger thought to contain a motif vital for antibody binding [45, 46]. Homology between the MS4A genes is observed in the transmembrane regions and this may suggest that the transmembrane regions may be most important for the function of these proteins [46].

Human CD20 expression commences at the early pre-B cell stage and is maintained until terminal differentiation into plasma cells [47]. The interaction of transcription factors to control CD20 expression is incompletely understood. However, it seems that CD20 expression is regulated by both B cell-specific transcription factors as well as other proteins that are not dependent on any specific stage of B cell development [46]. After expression, CD20 appears to reside in the cell membrane as either dimers or tetramers [48, 49]. It has been reported to localise with CD40, MHC class II and the B cell receptor for antigen (BCR) [50–52]. Both CD20 and BCR target to microvilli, which are specific membrane protrusions; it has been suggested that this may indicate a combined role for CD20 and the BCR in detecting antigen and subsequent signalling [53]. However, the precise function of CD20 is not known. There are only limited numbers of reagents to murine CD20 [37] and most information has been deduced from the use of mAbs to CD20 on human B cells and B cell lines. Different mAbs to CD20 produce a variety of different effects, including enhanced survival [54], activation and proliferation [55, 56], growth inhibition [57] and cell death [57–62]. Some of these different mAbs have been shown to induce a range of different calcium flux responses [48] and this may support the notion that CD20 may be involved in the generation and regulation of calcium transport triggered by other receptors. The release of calcium from intracellular stores, when combined with the influx of extracellular calcium *via* membrane channels leads to a rise in cytoplasmic calcium levels, which is required for B cell activation. Calcium flux is observed upon stimulation of the BCR and CD19 and this is reduced in CD20 knockout mice [37]. It is thought that CD20 may be the ion channel that can open to allow influx of calcium to replace depleted intracellular stores [63].

A variety of anti-CD20 mAbs have now been described. It has been suggested that these mAbs can be divided into two functional groups [62, 64, 65]. All appear to be equally effective at inducing ADCC but type I anti-CD20 mAbs, for example, rituximab, appear to induce much more potent complement-dependent cytotoxicity (CDC) mechanisms than type II anti-CD20 mAbs. The latter appear to induce more aggregation and apoptosis [66]. This functional distinction could have implications in the mechanism of action of different mAbs targeting CD20 with therapeutic consequences.

The contribution of various mechanisms of action of anti-CD20 mAbs in leading to B cell depletion is not completely understood. Four different possibilities exist: ADCC through effector cells, the involvement of complement (CDC), direct induction of cell death or growth inhibition and the stimulation of host-adaptive immunity [46]. The first three of these mechanisms are well supported by evidence but the last is less so.

The role of effector cells (such as NK cells and macrophages) bearing FcR has been demonstrated in both animal models [67] and clinical settings in both haematological malignancy and autoimmune disease. It is through the FcR that these cells interact with the Fc portions of Igs. There are three classes of FcR: FcRI (CD64), FcRII (CD32) and FcRIII (CD16). FcRIa, FcRIIa, FcRIIc and FcRIIIa are activating receptors, which when cross-linked lead eventually to cellular responses such as phagocytosis and ADCC [68]. FcRIIb is the only inhibitory FcR. Most cells express both activating and inhibitory receptors, although NK cells only express the former. In haematological disease, high-affinity antibody binding FcR allotypes correlate with improved response rates in follicular lymphoma treated with rituximab [69]. Human FcγRIIIa has polymorphisms at various positions; at position 158, this results in either valine (V) or phenylalanine (F) expression. The former has been associated with improved response to rituximab and greater depletion of B cells. In individuals with the FcRIIa-158 V/V genotype, increased CD16 expression on NK cells, rituximab binding and augmented rituximab-dependent ADCC were observed [70]. In systemic lupus erythematosus, the FcγRIIIa genotype has been also been related to the degree of B cell depletion after rituximab [71].

A variety of sources provide evidence for the role of complement mechanisms in CD20 immunotherapy. Complement is consumed during rituximab administration [72], thereby demonstrating that it is activated after therapy. Other experiments have demonstrated that the sensitivity of lymphoma cells to CDC *in vitro* is linked to their resistance to rituximab *in vitro* [73]. Systems in which complement has been removed, e.g. C1q-deficient mice, have also confirmed the role of CDC [74]. Beyond direct CDC, complement may stimulate opsonisation and may also promote the activity of effector cells. These mechanisms are being investigated [46].

Several factors affect the potency of CD20 in driving complement mechanisms on ligation with mAb. CD20 does not appear to be modulated *in vitro* after such ligation [75–77] and so complement activation is not adversely affected by downregulation of the initial stimulus. However, whether this lack of modulation still holds in *in vivo* situations is less clear [78]. The structure of CD20 is such that mAb is bound close to the plasma cell membrane and this is also the case for other potent CDC antigens, such as CD52 [79]. Another important factor appears to be the mobility and positioning of antibody-antigen complexes in the plasma cell membrane. This draws on the distinction made earlier between different types of mAb to anti-CD20. Thus, type I mAbs induce translocation into Triton X-100 (Tx-100)-insoluble lipid rafts and are much more potent at activating lytic complement than type II mAbs, which do not induce such lateral translocation of antibody–antigen complexes [62] (lipid rafts are highly ordered membrane domains abundant in glycosphingolipids, cholesterol and signalling proteins such as Src kinases).

Anti-CD20 mAbs can also trigger signals leading to cell death or cell arrest [58–62, 64]. In patients treated with rituximab for B cell chronic lymphocytic leukaemia (B-CLL), better clinical responses and more extensive depletion of CD20-positive cells correlated with apoptosis as detected by caspase-3 and -9 processing [80]. The redistribution of CD20 into lipid rafts may also be of importance. Changes in the regulation and activity of Src-like tyrosine kinases, which are present in high levels in that environment, may be important in CD20-induced apoptosis [59]. Other methods of inducing cell death not related to lipid rafts may also include a certain degree of CD20 cross-linking, which triggers apoptotic pathways, including caspase activation [81].

CD20 is an attractive target antigen for a number of reasons. The range of its B cell expression means that depletion of CD20-positive cells leaves the population of stem cells and plasma cells intact, ensuring eventual repletion of B cells as well as IgG levels, at least in the early stages of therapy. The lack of modulation of CD20 also makes it attractive for therapies that may work *via* CDC and ADCC. These mechanisms are also aided by the high expression of CD20 (100 000–200 000 copies on most B cell lines).

Future research into CD20 as a therapeutic target, particularly in autoimmune disease, may centre on the antigen itself as well as targeting antibodies. With regard to the former, questions have been raised about the the presence of a circulating soluble CD20 (cCD20), which may block the binding of rituximab to target cells. Such cCD20 has been detect in patients with non-Hodgkin's lymphoma (NHL) and CLL, and to a lesser extent in normal controls [82, 83]. This could also affect the measurement of rituximab concentrations in pharmocokinetic studies. Many of the factors influencing the pharmacokinetics of rituximab are incompletely understood. The initial choices of doses in haematology, which influenced to a degree those eventually selected in rheumatology, were mostly based on industrial considerations [84]. Relationships have been reported between rituximab exposure and tumour response [85] and progression-free survival [86]. In systemic lupus erythematosus (SLE), the degree of lymphocyte depletion correlated inversely with rituximab concentration 2 months after treatment [71]. In both haematological and autoimmune disease, rituximab pharmacokinetics can be described by a two-compartment model [84]. This suggests that the expression of CD20 (presumably higher in patients with tumours than in autoimmune disease) does not affect rituximab pharmacokinetics and exposure. Thus, the variation in exposure to rituximab is not yet fully explained but may relate to factors influencing the avidity of rituximab binding, e.g. FcR subtypes and polymorphisms therein, as described earlier.

With regard to anti-CD20 mAbs as a group, which antibody is optimal for the different disease processes it is not yet clear. Rituximab may work mainly *via* CDC mechanisms. Complement processes may be most effective in situations where the target cell is blood borne, allowing good access, i.e. type I anti-CD20 mAbs may be most suitable. Whether this applies to autoimmune rheumatic diseases is also not entirely clear and indeed in different diseases, such as RA and SLE, different mAbs may be useful. In the former, it remains to be seen how well rituximab affects B cells in compartments other than blood (e.g. synovium) and whether this is relevant to clinical outcomes; in the latter, which is often associated with complement deficiency or defects, mAbs that act by non-complement-dependent mechanisms may be more useful. Understanding the cell death pathway more completely will clearly be of key importance and recent advances demonstrating the possibility that CD20 may be a calcium channel may be of great relevance here. Parallel to this lies further work in developing different CD20 mAbs and studying the true differences in the mechanisms by which different mAbs induce cell death.

The BLyS system and its role in RA

The molecules of the B lymphocyte stimulator (BLyS) system seem to have a role at most, if not all, stages of B cell development. Two related TNF family ligands and three TNFR family receptors have been described that have complex and sometimes overlapping binding and function in B cell homeostasis (Fig. 2). The ligands are BLyS (also commonly called BAFF and TALL-1) [87–90] and 'a proliferation-inducing ligand' (APRIL, TALL-2) [91]. These interact with the receptors transmembrane activator and calcium modulator and cyclophylin ligand interactor (TACI, CD267), BLyS receptor 3 (BR3, BAFF-R, CD268) and B cell maturation antigen (BCMA, CD269). BLyS binds strongly to BR3, and more weakly to BCMA. APRIL binds strongly to BCMA. Both ligands bind TACI equally. The function of individual components has been elucidated in mouse and human studies.

The ligands BLyS and APRIL are closely homologous and, like other members of the TNF family, form active soluble homo- and heterotrimers [92–94]. Although BLyS is predominantly produced by monocytes, macrophages and dendritic cells [95], there is also evidence of production by neutrophils [96], and in some situations B cells [97, 98]. The release of BLyS by myeloid cells is stimulated by pro-inflammatory cytokines including interferon (IFN)-γ, TNF-α and IL-10 [95, 99].

Functions of BLyS

When mice are treated with BLyS, splenic B cell numbers are increased [87] due to increased B cell survival [100]. This is mediated by modifying the ratio of expression

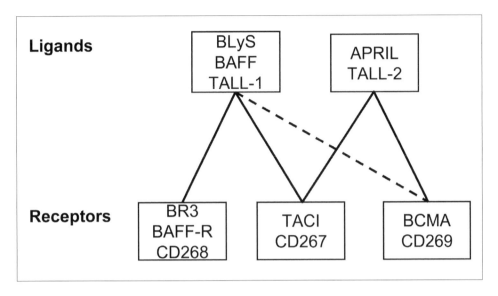

Figure 2
Binding between ligands and receptors of the BLyS/APRIL system. Note that BLyS and APRIL can also form homo- or heterotrimers that are biologically active (solid lines indicate stronger binding than the dashed line).

of pro- and anti-apoptotic molecules [101]. BLyS transgenic mice have expansion of mature B cells in peripheral blood, lymph nodes and spleen. They also develop hypergammaglobulinaemia, autoantibodies characteristic of human rheumatic diseases including RF, antibodies to nuclear antigens and cryoglobulins, and also renal immune complex deposition [102, 103]. In contrast, BLyS knockout mice have T1, but no T2, marginal zone or follicular B cells in the spleen or lymph nodes and no B2 in the peritoneum and markedly reduced Ig levels [104]. In contrast to mice, neither transitional nor plasma cells proliferate in response to BLyS in humans, but mature B cells and plasmablasts do [105]. As well as its role in B cell development and homeostasis, BLyS is important for normal B cell function in response to antigen. The production of BLyS by dendritic cells under stimulation by CD40L and IFN-γ has been shown to be crucial to induction of class-switch recombination following BCR engagement [106].

Functions of APRIL

The function of APRIL is less well understood. APRIL does not have equivalent *in vitro* B cell stimulatory properties [107, 108]. APRIL transgenic animals do not have an abnormal B cell phenotype or B cell-mediated autoimmune disease as occurs in BLyS transgenic animals [108]. APRIL knockout animals have apparently normal immune systems. APRIL does, however, form heterotrimers with BLyS in patients with autoimmune disease, and may have a role in modulating BLyS function [109].

BLyS and APRIL receptors

Human transitional, naïve, germinal centre and memory B cells in the spleen, peripheral blood and bone marrow all bind BLyS, but pro-, pre- and immature B cells do not. Human subsets of B cells differ in their expression of receptor for BLyS and APRIL, which determines their sensitivity to these ligands (Fig. 1).

In transitional and naïve B cells, only BR3 is expressed. Other BLyS receptors are only expressed on later subsets. Memory B cells bind BLyS by either BR3 or TACI, but not BCMA [110]. BLyS enhances their survival but not proliferation. After differentiating into plasmablasts, BR3 is down-regulated, and TACI and BCMA are expressed [111]; BLyS also enhances survival of these cells. Terminally differentiated plasma cells, however, do not express any BLyS receptor nor respond functionally [110].

BLyS system in RA

Serum levels of BLyS and APRIL are elevated in many rheumatic diseases [112] and in RA higher levels in synovium confirm that there is local production here [34]. The production and function of BLyS system molecules differs in synovia with different lymphoid architecture [113]. While macrophages produce BLyS in all types of rheumatoid synovitis, APRIL production (by dendritic cells) is greatest when there is a germinal centre-like pattern, lower in aggregate synovitis and lowest in diffuse synovitis. Similarly, although BR3 and TACI are expressed equally in all tissues, the relatively APRIL-specific receptor BCMA is expressed more strongly in germinal centre or aggregate synovitis than in diffuse synovitis. Treatment of a human synovium-SCID mouse chimera with TACI-Fc inhibits Ig and IFN-γ production in germinal centre synovitis; it also destroys the germinal centres themselves and reduces B, T and follicular dendritic cell numbers. In diffuse and aggregate synovitis however, Ig production is not affected by treatment with TACI-Fc, and IFN-γ is actually increased. Thus, the roles of BLyS and APRIL in the rheumatoid synovium are complex and may vary according to disease subtype.

Agents that target CD20

Rituximab

Data regarding the clinical efficacy and safety of rituximab are largely derived from three double-blind, randomised, placebo-controlled studies, each of which demonstrated superiority to methotrexate in control of disease activity [114–116].

In RA, rituximab has been given as a fixed dose (i.e. not determined by surface area) as two infusions, separated by 14 days. Each infusion comprised 500 mg or 1 g. The first Phase IIa study [115] indicated that the combination of two infusions of 1 g rituximab with methotrexate (10–25 mg weekly) was significantly better than methotrexate alone in allowing patients to achieve American College of Rheumatology (ACR) 50 response 6 months after therapy (43% vs 13%, $p = 0.005$). The REFLEX study [114], which investigated patients who had inadequate response to anti-TNF therapy, also demonstrated that use of two infusions of 1 g rituximab with methotrexate was superior to methotrexate alone in achieving ACR20 response after 6 months (51% vs 18% respectively, $p < 0.0001$).

The Phase IIa study [115] demonstrated that the combination of rituximab with either methotrexate or cyclophosphamide was significantly better than methotrexate monotherapy in achieving ACR50 response at 24 weeks and this was maintained at 48 weeks. However, the combination of rituximab and methotrexate was the only one to demonstrate significantly higher frequency of ACR70 response at 24 weeks compared to the control group and has become the standard combination in subsequent studies. Although rituximab monotherapy was significantly better than methotrexate monotherapy at 24 weeks ($p < 0.002$), this improvement was no longer statistically significant at week 48. How methotrexate prolongs the benefits of rituximab therapy is not yet fully understood. The concomitant use of methotrexate does not appear to affect the duration of B cell depletion [117]. There are case series that suggest that other disease-modifying anti-rheumatic drugs (DMARDs) may be safe and effective in combination with rituximab [118], although further studies are required with such agents.

The initial use of rituximab in RA drew considerably on experience from haematology and so rituximab infusions were preceded by intravenous steroids and a course of oral steroids was prescribed between infusions. The DANCER study showed that intravenous steroids significantly reduced the incidence and severity of acute infusion reactions but neither intravenous nor oral steroids appear to have any significant effect on efficacy at 24 weeks [116].

Infusion reactions and infections

A pooled analysis with regard to safety has recently been undertaken of 1039 patients in the rituximab RA treatment programme [119]. Over 3 years, rituximab and placebo-treated patients experienced similar numbers of serious and non-serious adverse events. Infusion-related events included , pruritus, urticaria, pyrexia, throat irritation and hypertension. In rituximab-treated patients it is thought that these reactions are due to cytokine release following B cell lysis. Such symptoms affected approximately 15% of rituximab-treated patients but it should be noted that the incidence following the second infusion was much lower. This is presumably because B cell numbers are already significantly reduced by the first infusion. A small number of patients (< 1%) experienced a serious infusion reaction (anaphylaxis, bronchospasm, etc). Patients were less likely to experience an infusion reaction if they received premedication with intravenous corticosteroids and if they received the lower (500 mg) dose of rituximab. The incidence of infusion reactions declined with subsequent courses of rituximab (as did all adverse events).

With regard to infections, serious infections (i.e. requiring intravenous antibiotics or classified as serious adverse event) occurred in 7% of rituximab patients compared with 3% of placebo patients. The overall rate of serious infections was 5.03 events/100 patient years. Prolonged peripheral B cell depletion was not associated with an increased rate of such infections.

In the REFLEX study, differences in ACR20 responses over placebo were significant for both seropositive and seronegative patients [114]. Of RF-positive patients, 20% became negative following therapy but despite this these patients displayed similar clinical responses following subsequent courses of therapy. Seronegative patients that were positive for anti-citrullinated protein antibodies had similar responses to seropositive patients, in contrast to patients negative for both antibodies who had lower response rates [120].

B cell depletion, reconstitution and retreatment

Conventional flow cytometry analysis suggests that there is nearly complete depletion of B cells from the peripheral blood following rituximab therapy in all patients. This being so, there would seem to be no correlation between B cell depletion and clinical response. However, highly sensitive flow cytometric techniques, developed to assess disease response in haematological malignancies and known as minimal residual disease flow cytometry (MRD Flow), has given early indication that B cell depletion may not be as complete as initially thought [121]. In terms of B cell reconstitution, data from the initial Phase IIa study are extensive, with patients now being followed for up to 2 years. The data suggest that B cells begin to rise between 6 and 12 months after treatment in most patients. Clinical deterioration appears to occur between 6 and 12 months after the initial infusion [122]. The relationship between the return of B cells and that of disease activity is not clearly defined. In individual patients, the phenomenon of B cell counts rising with concurrent disease relapse is well observed. However, some patients have demonstrated return of B cells in peripheral blood without relapse. In an open label study, in patients in whom return of disease was associated with return of B cells, there was a higher proportion of memory B cells compared to those who underwent B cell reconstitution without relapse [123].

In the clinical trial setting, patients with relapse of disease following initial response were identified (on clinical grounds: numbers of swollen and tender joint counts) and entered into retreatment protocols. Most patients who received retreatment did so when their B cell levels, although rising, were lower than before their first exposure to rituximab. Retreatment has generally been well tolerated and effective [124, 125]. The median time between the first and second course of rituximab was 30.9 weeks for patients who had received prior anti-TNF therapy and 36.7 weeks in those who had not had prior anti-TNF therapy. The median time between second and third courses was 30.1 weeks for the former and 43.0 for the latter [126]. Patients generally received retreatment at levels of disease activity that were lower than at baseline, hence ACR responses (from original baseline) following the second course were better than those seen after the first course and similar when compared with the course baseline. When EULAR responses are considered, more patients had good responses and entered low disease states or remission after the second course than the first.

With regards to the safety of repeat courses of rituximab, data are available from 570 patients who have had a second course and 191 who have had a third [124, 125]. For patients who had received anti-TNF previously, the median time to retreatment with a second course was 30.9 weeks and between second and third courses was 30.1 weeks. These times for patients who had not previously received anti-TNF were 36.7 and 40.3 weeks, respectively [126]. The rate of infections (including those deemed serious) did not change significantly with repeated therapy. The incidence of infusion reactions fell with the second and subsequent courses of therapy. Ig levels fell with repeated therapy, and after the third course, 23.5% of patients had lower levels of IgM than normal. However, the rate of serious infections before and after detection of low IgM did not change significantly (5.1 vs 5.9 per 100 patient years, respectively).

A number of patients from the initial study (45% of those receiving one cycle of rituximab with continuing methotrexate) did not have worsening of disease that warranted additional therapy, and improvements in physical function were present at 2 years after therapy [127, 128]. In the small number of patients who had prolonged B cell depletion for up to 2 years after therapy, there was no increase in the rates of infection. The median duration of B cell depletion in patients who experienced infections was similar to that in patients who did not have infections.

Ocrelizumab

Ocrelizumab is a humanised anti-CD20 antibody [129]. A dose-ranging Phase I/II trial has shown B cell depletion at all doses, with earlier repletion at lower doses.

Clinical responses were observed across the doses that were improved over placebo. Further dose-finding studies are being conducted [130].

Ofatumumab

Ofatumumab (HuMax-CD20) is a fully humanised anti-CD20 antibody being studied in follicular NHL and CLL (where it is in late stage development) and has been used in a dose-ranging Phase I/II trial in RA. It has been well tolerated and 63% of the patients receiving active drug (20/32) achieved ACR50 response compared with none of the placebo patients [131].

Agents that target the BLyS system

Belimumab

Belimumab (Lymphostat-B) is a humanised mAb to BLyS that inhibits its activity. Results from a Phase II double-blind placebo-controlled trial in RA have been reported in which 283 patients were randomised to belimumab or placebo. The primary endpoint of ACR20 was met in 29% of patients in all active treatment arms compared to 16% in placebo, and there was no dose response [132]. Levels of RF and Igs were reduced, as were total B cells, but there was an increase in memory B cells [133]. Plasma cell numbers were not affected.

These disappointing results may be because of overlapping signalling between other members of the BLyS system or the failure of belimumab to block BLyS/APRIL heterotrimers. Trials in SLE, however, continue and have shown some evidence of efficacy.

Atacicept

Atacicept (TACI-Ig) is a fusion protein consisting of the APRIL/BLyS-binding extracellular portion of the TACI molecule fused to the Fc portion of human IgG1, and binds both BLyS and APRIL as well as homo- and heterotrimers [109]. It is in development for the treatment of RA, SLE, and B-CLL. In animal studies it resulted in a reduction of mature B cells in peripheral blood and lymphoid organs, as well as a reduction in circulating Ig [134]. In a Phase Ib trial in human RA there was a reduction in RF and anti-citrullinated protein antibodies as well as serum Ig [135]. The drug had a good safety profile and penetrated inflamed joints, and there was a trend to improved disease activity.

BR3-Fc

BR3-Fc is a fusion protein consisting of two polypeptide chains linked by disulphide bonds, with sequences from the extracellular domain of the human BAFF receptor, BR3, and the Fc domain of human IgG [136]. Binding of BAFF to soluble BR3-Fc protein blocks binding to cell-bound BR3, inhibiting B cell activation and resulting in apoptosis. BR3-Fc reduced B cells in peripheral blood and marginal zone of lymphoid tissue as well as follicular B cells in animal studies. Results are available from a Phase I study in patients with RA [137]. Peripheral blood B cells were reduced and there was no serious toxicity.

Conclusions

Scientific understanding of the role of B cells in RA has advanced significantly in recent years. Although RF has long been thought to be of key importance, the mechanisms by which RF-producing B cells cause directed joint disease as well as sustained T cell involvement through the continuing presentation of other autoantibody complexes are now better understood. A vicious circle of autoimmunity develops in RA. B cell depletion with rituximab has been partially successful at breaking this circle and its clinical efficacy has also shed light on the pathological mechanisms in RA.

CD20 has been a serendipitous target in RA, in terms of its restricted expression, allowing B cells to be depleted without directly affecting stem cells or plasma cells. The action of rituximab *via* CD20 is generally effective for depletion, although further understanding of the mechanism of this action might allow enhancement of rituximabs effects, either on an individual basis or based on disease type. In parallel with this, further understanding of the pharmacokinetics of rituximab could allow better dosing regimens. Further study of CD20 will help in the development of other monoclonal therapies directed against this.

Clinical experience with rituximab has indicated that peripheral B cell depletion is associated with reduction of disease activity in RA. The factors that influence response to rituximab – both initially and in terms of duration of response – are unclear. It is now also recognised that in the vast majority of patients, both peripheral B cells and disease activity return. However, the temporal and pathological links between these two phenomena are incompletely understood. It may be that the absolute depth of B cell depletion is critical. Alternatively, a specific B cell subset might need to be eliminated (or be prevented from returning) to achieve durable clinical response. The effect of rituximab on B cells in compartments other than peripheral blood is still being investigated. The ultimate therapeutic effect of B cell depletion may lie in synovium or lymphoid tissue rather than in peripheral blood.

As B cell depletion becomes more commonplace, attention has turned to methods of B cell modulation. Targets such as BLyS and APRIL are being more closely studied and some initial clinical data are available. The variable results so far may indicate that these therapies may act best as adjunct, possibly to depleting agents. However, their role needs to be investigated further, particularly as the long-term consequences of repeated B cell depletion remain unclear, although hypogammaglobulinaemia is recognised.

The discovery that B cell targeting is effective in RA is of great importance, given the overall disease burden and the limited therapeutic options available to many patients. Scientific understanding of B cell pathology as well as therapeutic mechanisms is greatly advancing our understanding of RA pathogenesis. Identification of B cell targets and development of appropriate therapy is already changing treatment paradigms in RA and is likely to continue do so.

References

1   Hoyer BF, Manz RA, Radbruch A, Hiepe F (2005) Long-lived plasma cells and their contribution to autoimmunity. *Ann N Y Acad Sci* 1050: 124–133

2   Kaplan RA, Curd JG, Deheer DH, Carson DA, Pangburn MK, Muller-Eberhard HJ, Vaughan JH (1980) Metabolism of C4 and factor B in rheumatoid arthritis. Relation to rheumatoid factor. *Arthritis Rheum* 23: 911–920

3   Ruddy S, Britton MC, Schur PH, Austen KF (1969) Complement components in synovial fluid: Activation and fixation in seropositive rheumatoid arthritis. *Ann N Y Acad Sci* 168: 161–172

4   Bhatia A, Blades S, Cambridge G, Edwards JC (1998) Differential distribution of Fc gamma RIIIa in normal human tissues and co-localization with DAF and fibrillin-1: Implications for immunological microenvironments. *Immunology* 94: 56–63

5   Abrahams VM, Cambridge G, Lydyard PM, Edwards JC (2000) Induction of tumor necrosis factor alpha production by adhered human monocytes: A key role for Fcgamma receptor type IIIa in rheumatoid arthritis. *Arthritis Rheum* 43: 608–616

6   Ji H, Ohmura K, Mahmood U, Lee DM, Hofhuis FM, Boackle SA, Takahashi K, Holers VM, Walport M, Gerard C et al (2002) Arthritis critically dependent on innate immune system players. *Immunity* 16: 157–168

7   Neumann E, Barnum SR, Tarner IH, Echols J, Fleck M, Judex M, Kullmann F, Mountz JD, Scholmerich J, Gay S et al (2002) Local production of complement proteins in rheumatoid arthritis synovium. *Arthritis Rheum* 46: 934–945

8   Fehr K, Velvart M, Rauber M, Knopfel M, Baici A, Salgam P, Boni A (1981) Production of agglutinators and rheumatoid factors in plasma cells of rheumatoid and nonrheumatoid synovial tissues. *Arthritis Rheum* 24: 510–519

9   Van Snick JL, Van Roost E, Markowetz B, Cambiaso CL, Masson PL (1978) Enhance-

ment by IgM rheumatoid factor of *in vitro* ingestion by macrophages and *in vivo* clearance of aggregated IgG or antigen-antibody complexes. *Eur J Immunol* 8: 279–285

10 Devey ME, Hogben DN (1987) The effect of rheumatoid factor on the clearance of endogenous immune complexes formed in low-affinity mice during the induction of immune complex disease. *Int Arch Allergy Appl Immunol* 83: 206–209

11 Roosnek E, Lanzavecchia A (1991) Efficient and selective presentation of antigen-antibody complexes by rheumatoid factor B cells. *J Exp Med* 173: 487–489

12 Kyburz D, Corr M, Brinson DC, Von Damm A, Tighe H, Carson DA (1999) Human rheumatoid factor production is dependent on CD40 signaling and autoantigen. *J Immunol* 163: 3116–3122

13 Mannik M, Nardella FA, Sasso EH (1988) Rheumatoid factors in immune complexes of patients with rheumatoid arthritis. *Springer Semin Immunopathol* 10: 215–230

14 Edwards JC, Cambridge G (1998) Rheumatoid arthritis: The predictable effect of small immune complexes in which antibody is also antigen. *Br J Rheumatol* 37: 126–130

15 Schroder AE, Greiner A, Seyfert C, Berek C (1996) Differentiation of B cells in the non-lymphoid tissue of the synovial membrane of patients with rheumatoid arthritis. *Proc Natl Acad Sci USA* 93: 221–225

16 Takemura S, Braun A, Crowson C, Kurtin PJ, Cofield RH, O'Fallon WM, Goronzy JJ, Weyand CM (2001) Lymphoid neogenesis in rheumatoid synovitis. *J Immunol* 167: 1072–1080

17 Weyand CM, Goronzy JJ (2003) Ectopic germinal center formation in rheumatoid synovitis. *Ann N Y Acad Sci* 987: 140–149

18 Thurlings RM, Wijbrandts CA, Mebius RE, Cantaert T, Dinant H, van der Pouw-Kraan TCTM, Baeten D, Tak PP (2008) Synovial lymphoid neogenesis does not define a specific clinical rheumatoid arthritis phenotype. *Arthritis Rheum* 58: 1582–1589

19 Tak PP, Smeets TJM, Daha MR, Kluin PM, Meijers KAE, Brand R, Meinders AE, Breedveld FC (1997) Analysis of the synovial cellular infiltrate in early rheumatoid synovial tissue in relation to local disease activity. *Arthritis Rheum* 40: 217–225

20 Weyand CM, Seyler TM, Goronzy JJ (2005) B cells in rheumatoid synovitis. *Arthritis Res Ther* 7 Suppl 3: S9–12

21 Voswinkel J, Weisgerber K, Pfreundschuh M, Gause A (2001) B lymphocyte involvement in ankylosing spondylitis: The heavy chain variable segment gene repertoire of B lymphocytes from germinal center-like foci in the synovial membrane indicates antigen selection. *Arthritis Res* 3: 189–195

22 van Kuijk AW, Reinders-Blankert P, Smeets TJ, Dijkmans BA, Tak PP (2006) Detailed analysis of the cell infiltrate and the expression of mediators of synovial inflammation and joint destruction in the synovium of patients with psoriatic arthritis: Implications for therapy. *Ann Rheum Dis* 65: 1551–1557

23 Kraan MC, Haringman JJ, Post WJ, Versendaal J, Breedveld FC, Tak PP (1999) Immunohistological analysis of synovial tissue for differential diagnosis in early arthritis. *Rheumatology* 38: 1074–1080

24 Clausen BE, Bridges SL Jr, Lavelle JC, Fowler PG, Gay S, Koopman WJ, Schroeder HW

Jr (1998) Clonally-related immunoglobulin VH domains and nonrandom use of DH gene segments in rheumatoid arthritis synovium. *Mol Med* 4: 240–257

25 Gause A, Gundlach K, Carbon G, Daus H, Trumper L, Pfreundschuh M (1997) Analysis of VH gene rearrangements from synovial B cells of patients with rheumatoid arthritis reveals infiltration of the synovial membrane by memory B cells. *Rheumatol Int* 17: 145–150

26 Goronzy JJ, Zettl A, Weyand CM (1998) T cell receptor repertoire in rheumatoid arthritis. *Int Rev Immunol* 17: 339–363

27 Hauser AE, Debes GF, Arce S, Cassese G, Hamann A, Radbruch A, Manz RA (2002) Chemotactic responsiveness toward ligands for CXCR3 and CXCR4 is regulated on plasma blasts during the time course of a memory immune response. *J Immunol* 169: 1277–1282

28 Tsubaki T, Takegawa S, Hanamoto H, Arita N, Kamogawa J, Yamamoto H, Takubo N, Nakata S, Yamada K, Yamamoto S et al (2005) Accumulation of plasma cells expressing CXCR3 in the synovial sublining regions of early rheumatoid arthritis in association with production of Mig/CXCL9 by synovial fibroblasts. *Clin Exp Immunol* 141: 363–371

29 Kamogawa J TN, Arita N, Okada K, Yamamoto H, Yamamoto S, Nose M (2000) Histopathological characteristics of early rheumatoid arthritis: A case one month after clinical onset. *Mod Rheumatol* 10: 272–275

30 Dass S BC, Vital EM, Reece RJ, Rawstron AC, Ponchel F, Emery P (2007) Reduction in synovial B cells after rituximab in RA predicts clinical response. *Ann Rheum Dis* 66: 90

31 Vos K, Thurlings RM, Wijbrandts CA, van Schaardenburg D, Gerlag DM, Tak PP (2007) Early effects of rituximab on the synovial cell infiltrate in patients with rheumatoid arthritis. *Arthritis Rheum* 56: 772–778

32 Kavanaugh A, Rosengren S, Lee SJ, Hammaker D, Firestein GS, Kalunian K, Wei N, Boyle DL (2008) Assessment of rituximab's immunomodulatory synovial effects (the ARISE trial). I: Clinical and synovial biomarker results. *Ann Rheum Dis* 67: 402–408

33 Dechanet J, Merville P, Durand I, Banchereau J, Miossec P (1995) The ability of synoviocytes to support terminal differentiation of activated B cells may explain plasma cell accumulation in rheumatoid synovium. *J Clin Invest* 95: 456–463

34 Tan SM, Xu D, Roschke V, Perry JW, Arkfeld DG, Ehresmann GR, Migone TS, Hilbert DM, Stohl W (2003) Local production of B lymphocyte stimulator protein and APRIL in arthritic joints of patients with inflammatory arthritis. *Arthritis Rheum* 48: 982–992

35 Gong Q, Ou Q, Ye S, Lee WP, Cornelius J, Diehl L, Lin WY, Hu Z, Lu Y, Chen Y et al (2005) Importance of cellular microenvironment and circulatory dynamics in B cell immunotherapy. *J Immunol* 174: 817–826

36 Thurlings RM, Vos K, Wijbrandts CA, Zwinderman A, Gerlag DM, Tak PP (2008) Synovial tissue response to rituximab: Mechanism of action and identification of biomarkers of response. *Ann Rheum Dis* 67: 917–925

37   Uchida J, Lee Y, Hasegawa M, Liang Y, Bradney A, Oliver JA, Bowen K, Steeber DA, Haas KM, Poe JC et al (2004) Mouse CD20 expression and function. *Int Immunol* 16: 119–129

38   O'Keefe TL, Williams GT, Davies SL, Neuberger MS (1998) Mice carrying a CD20 gene disruption. *Immunogenetics* 48: 125–132

39   Tedder TF, Klejman G, Disteche CM, Adler DA, Schlossman SF, Saito H (1988) Cloning of a complementary DNA encoding a new mouse B lymphocyte differentiation antigen, homologous to the human B1 (CD20) antigen, and localization of the gene to chromosome 19. *J Immunol* 141: 4388–4394

40   Tedder TF, Streuli M, Schlossman SF, Saito H (1988) Isolation and structure of a cDNA encoding the B1 (CD20) cell-surface antigen of human B lymphocytes. *Proc Natl Acad Sci USA* 85: 208–212

41   Stamenkovic I, Seed B (1988) Analysis of two cDNA clones encoding the B lymphocyte antigen CD20 (B1, Bp35), a type III integral membrane protein. *J Exp Med* 167: 1975–1980

42   Einfeld DA, Brown JP, Valentine MA, Clark EA, Ledbetter JA (1988) Molecular cloning of the human B cell CD20 receptor predicts a hydrophobic protein with multiple transmembrane domains. *EMBO J* 7: 711–717

43   Liang Y, Buckley TR, Tu L, Langdon SD, Tedder TF (2001) Structural organization of the human MS4A gene cluster on Chromosome 11q12. *Immunogenetics* 53: 357–368

44   Liang Y, Tedder TF (2001) Identification of a CD20-, FcepsilonRIbeta-, and HTm4-related gene family: Sixteen new MS4A family members expressed in human and mouse. *Genomics* 72: 119–127

45   Polyak MJ, Tailor SH, Deans JP (1998) Identification of a cytoplasmic region of CD20 required for its redistribution to a detergent-insoluble membrane compartment. *J Immunol* 161: 3242–3248

46   Cragg MS, Walshe CA, Ivanov AO, Glennie MJ (2005) The biology of CD20 and its potential as a target for mAb therapy. *Current Direct Autoimmun* 8: 140–174

47   Tedder TF, Engel P (1994) CD20: A regulator of cell-cycle progression of B lymphocytes. *Immunol Today* 15: 450–454

48   Bubien JK, Zhou LJ, Bell PD, Frizzell RA, Tedder TF (1993) Transfection of the CD20 cell surface molecule into ectopic cell types generates a Ca^{2+} conductance found constitutively in B lymphocytes. *J Cell Biol* 121: 1121–1132

49   Polyak MJ, Deans JP (2002) Alanine-170 and proline-172 are critical determinants for extracellular CD20 epitopes; heterogeneity in the fine specificity of CD20 monoclonal antibodies is defined by additional requirements imposed by both amino acid sequence and quaternary structure. *Blood* 99: 3256–3262

50   Leveille C, R AL-D, Mourad W (1999) CD20 is physically and functionally coupled to MHC class II and CD40 on human B cell lines. *Eur J Immunol* 29: 65–74

51   Szollosi J, Horejsi V, Bene L, Angelisova P, Damjanovich S (1996) Supramolecular complexes of MHC class I, MHC class II, CD20, and tetraspan molecules (CD53, CD81, and CD82) at the surface of a B cell line JY. *J Immunol* 157: 2939–2946

52   Petrie RJ, Deans JP (2002) Colocalization of the B cell receptor and CD20 followed by activation-dependent dissociation in distinct lipid rafts. *J Immunol* 169: 2886–2891

53   Li H, Ayer LM, Polyak MJ, Mutch CM, Petrie RJ, Gauthier L, Shariat N, Hendzel MJ, Shaw AR, Patel KD et al (2004) The CD20 calcium channel is localized to microvilli and constitutively associated with membrane rafts: Antibody binding increases the affinity of the association through an epitope-dependent cross-linking-independent mechanism. *J Biol Chem* 279: 19893–19901

54   Holder M, Grafton G, MacDonald I, Finney M, Gordon J (1995) Engagement of CD20 suppresses apoptosis in germinal center B cells. *Eur J Immunol* 25: 3160–3164

55   Smeland EB, Beiske K, Ek B, Watt R, Pfeifer-Ohlsson S, Blomhoff HK, Godal T, Ohlsson R (1987) Regulation of c-myc transcription and protein expression during activation of normal human B cells. *Exp Cell Res* 172: 101–109

56   Clark EA, Shu G (1987) Activation of human B cell proliferation through surface Bp35 (CD20) polypeptides or immunoglobulin receptors. *J Immunol* 138: 720–725

57   Tedder TF, Forsgren A, Boyd AW, Nadler LM, Schlossman SF (1986) Antibodies reactive with the B1 molecule inhibit cell cycle progression but not activation of human B lymphocytes. *Eur J Immunol* 16: 881–887

58   Shan D, Ledbetter JA, Press OW (2000) Signaling events involved in anti-CD20-induced apoptosis of malignant human B cells. *Cancer Immunol Immunother* 48: 673–683

59   Hofmeister JK, Cooney D, Coggeshall KM (2000) Clustered CD20 induced apoptosis: src-family kinase, the proximal regulator of tyrosine phosphorylation, calcium influx, and caspase 3-dependent apoptosis. *Blood Cells Mol Dis* 26: 133–143

60   Pedersen IM, Buhl AM, Klausen P, Geisler CH, Jurlander J (2002) The chimeric anti-CD20 antibody rituximab induces apoptosis in B-cell chronic lymphocytic leukemia cells through a p38 mitogen activated protein-kinase-dependent mechanism. *Blood* 99: 1314–1319

61   Cardarelli PM, Quinn M, Buckman D, Fang Y, Colcher D, King DJ, Bebbington C, Yarranton G (2002) Binding to CD20 by anti-B1 antibody or F(ab')(2) is sufficient for induction of apoptosis in B-cell lines. *Cancer Immunol Immunother* 51: 15–24

62   Cragg MS, Morgan SM, Chan HT, Morgan BP, Filatov AV, Johnson PW, French RR, Glennie MJ (2003) Complement-mediated lysis by anti-CD20 mAb correlates with segregation into lipid rafts. *Blood* 101: 1045–1052

63   Li H, Ayer LM, Lytton J, Deans JP (2003) Store-operated cation entry mediated by CD20 in membrane rafts. *J Biol Chem* 278: 42427–42434

64   Cragg MS AA, O'Brien L, Tutt A, Chan HTC, Anderson VA, Glennie MJ (2002) Opposing properties of CD20 mAb. In: *Leukocyte typing VII*. Oxford University Press, Oxford, 95–97

65   Cragg MS, Glennie MJ (2004) Antibody specificity controls *in vivo* effector mechanisms of anti-CD20 reagents. *Blood* 103: 2738–2743

66   Chan HT, Hughes D, French RR, Tutt AL, Walshe CA, Teeling JL, Glennie MJ, Cragg MS (2003) CD20-induced lymphoma cell death is independent of both caspases and its redistribution into Triton X-100 insoluble membrane rafts. *Cancer Res* 63: 5480–5489

67 Clynes RA, Towers TL, Presta LG, Ravetch JV (2000) Inhibitory Fc receptors modulate *in vivo* cytoxicity against tumor targets. *Nat Med* 6: 443–446

68 Cartron G, Watier H, Golay J, Solal-Celigny P (2004) From the bench to the bedside: Ways to improve rituximab efficacy. *Blood* 104: 2635–2642

69 Cartron G, Dacheux L, Salles G, Solal-Celigny P, Bardos P, Colombat P, Watier H (2002) Therapeutic activity of humanized anti-CD20 monoclonal antibody and polymorphism in IgG Fc receptor FcgammaRIIIa gene. *Blood* 99: 754–758

70 Hatjiharissi E, Xu L, Santos DD, Hunter ZR, Ciccarelli BT, Verselis S, Modica M, Cao Y, Manning RJ, Leleu X et al (2007) Increased natural killer cell expression of CD16, and augmented binding and ADCC activity to rituximab among individuals expressing the FcgammaRIIIa-158 V/V and V/F polymorphism. *Blood* 110: 2561–2564

71 Anolik JH, Campbell D, Felgar RE, Young F, Sanz I, Rosenblatt J, Looney RJ (2003) The relationship of FcgammaRIIIa genotype to degree of B cell depletion by rituximab in the treatment of systemic lupus erythematosus. *Arthritis Rheum* 48: 455–459

72 van der Kolk LE, Grillo-Lopez AJ, Baars JW, Hack CE, van Oers MH (2001) Complement activation plays a key role in the side-effects of rituximab treatment. *Br J Haematol* 115: 807–811

73 Manches O, Lui G, Chaperot L, Gressin R, Molens JP, Jacob MC, Sotto JJ, Leroux D, Bensa JC, Plumas J (2003) *In vitro* mechanisms of action of rituximab on primary non-Hodgkin lymphomas. *Blood* 101: 949–954

74 Di Gaetano N, Cittera E, Nota R, Vecchi A, Grieco V, Scanziani E, Botto M, Introna M, Golay J (2003) Complement activation determines the therapeutic activity of rituximab *in vivo*. *J Immunol* 171: 1581–1587

75 Press OW, Howell-Clark J, Anderson S, Bernstein I (1994) Retention of B-cell-specific monoclonal antibodies by human lymphoma cells. *Blood* 83: 1390–1397

76 Press OW, Farr AG, Borroz KI, Anderson SK, Martin PJ (1989) Endocytosis and degradation of monoclonal antibodies targeting human B-cell malignancies. *Cancer Res* 49: 4906–4912

77 Vervoordeldonk SF, Merle PA, van Leeuwen EF, von dem Borne AE, Slaper-Cortenbach IC (1994) Preclinical studies with radiolabeled monoclonal antibodies for treatment of patients with B-cell malignancies. *Cancer* 73: 1006–1011

78 Pickartz T, Ringel F, Wedde M, Renz H, Klein A, von Neuhoff N, Dreger P, Kreuzer KA, Schmidt CA, Srock S et al (2001) Selection of B-cell chronic lymphocytic leukemia cell variants by therapy with anti-CD20 monoclonal antibody rituximab. *Exp Hematol* 29: 1410–1416

79 Xia MQ, Hale G, Waldmann H (1993) Efficient complement-mediated lysis of cells containing the CAMPATH-1 (CDw52) antigen. *Mol Immunol* 30: 1089–1096

80 Byrd JC, Kitada S, Flinn IW, Aron JL, Pearson M, Lucas D, Reed JC (2002) The mechanism of tumor cell clearance by rituximab *in vivo* in patients with B-cell chronic lymphocytic leukemia: Evidence of caspase activation and apoptosis induction. *Blood* 99: 1038–1043

81 Rowan W, Tite J, Topley P, Brett SJ (1998) Cross-linking of the CAMPATH-1 antigen

(CD52) mediates growth inhibition in human B- and T-lymphoma cell lines, and subsequent emergence of CD52-deficient cells. *Immunology* 95: 427–436

82   Giles FJ, Vose JM, Do KA, Johnson MM, Manshouri T, Bociek G, Bierman PJ, O'Brien SM, Keating MJ, Kantarjian HM et al (2003) Circulating CD20 and CD52 in patients with non-Hodgkin's lymphoma or Hodgkin's disease. *Br J Haematol* 123: 850–857

83   Manshouri T, Do KA, Wang X, Giles FJ, O'Brien SM, Saffer H, Thomas D, Jilani I, Kantarjian HM, Keating MJ et al (2003) Circulating CD20 is detectable in the plasma of patients with chronic lymphocytic leukemia and is of prognostic significance. *Blood* 101: 2507–2513

84   Cartron G, Blasco H, Paintaud G, Watier H, Le Guellec C (2007) Pharmacokinetics of rituximab and its clinical use: Thought for the best use? *Crit Rev Oncol Hematol* 62: 43–52

85   Berinstein NL, Grillo-Lopez AJ, White CA, Bence-Bruckler I, Maloney D, Czuczman M, Green D, Rosenberg J, McLaughlin P, Shen D (1998) Association of serum Rituximab (IDEC-C2B8) concentration and anti-tumor response in the treatment of recurrent low-grade or follicular non-Hodgkin's lymphoma. *Ann Oncol* 9: 995–1001

86   Igarashi T, Kobayashi Y, Ogura M, Kinoshita T, Ohtsu T, Sasaki Y, Morishima Y, Murate T, Kasai M, Uike N et al (2002) Factors affecting toxicity, response and progression-free survival in relapsed patients with indolent B-cell lymphoma and mantle cell lymphoma treated with rituximab: A Japanese phase II study. *Ann Oncol* 13: 928–943

87   Moore PA, Belvedere O, Orr A, Pieri K, LaFleur DW, Feng P, Soppet D, Charters M, Gentz R, Parmelee D et al (1999) BLyS: Member of the tumor necrosis factor family and B lymphocyte stimulator. *Science* 285: 260–263

88   Mukhopadhyay A, Ni J, Zhai Y, Yu GL, Aggarwal BB (1999) Identification and characterization of a novel cytokine, THANK, a TNF homologue that activates apoptosis, nuclear factor-kappaB, and c-Jun NH2-terminal kinase. *J Biol Chem* 274: 15978–15981

89   Schneider P, MacKay F, Steiner V, Hofmann K, Bodmer JL, Holler N, Ambrose C, Lawton P, Bixler S, Acha-Orbea H et al (1999) BAFF, a novel ligand of the tumor necrosis factor family, stimulates B cell growth. *J Exp Med* 189: 1747–1756

90   Shu HB, Hu WH, Johnson H (1999) TALL-1 is a novel member of the TNF family that is down-regulated by mitogens. *J Leukoc Biol* 65: 680–683

91   Hahne M, Kataoka T, Schroter M, Hofmann K, Irmler M, Bodmer JL, Schneider P, Bornand T, Holler N, French LE et al (1998) APRIL, a new ligand of the tumor necrosis factor family, stimulates tumor cell growth. *J Exp Med* 188: 1185–1190

92   Liu Y, Hong X, Kappler J, Jiang L, Zhang R, Xu L, Pan CH, Martin WE, Murphy RC, Shu HB et al (2003) Ligand-receptor binding revealed by the TNF family member TALL-1. *Nature* 423: 49–56

93   Liu Y, Xu L, Opalka N, Kappler J, Shu HB, Zhang G (2002) Crystal structure of sTALL-1 reveals a virus-like assembly of TNF family ligands. *Cell* 108: 383–394

94   Zhukovsky EA, Lee JO, Villegas M, Chan C, Chu S, Mroske C (2004) TNF ligands: Is TALL-1 a trimer or a virus-like cluster? *Nature* 427: 413–414; discussion 414

95   Nardelli B, Belvedere O, Roschke V, Moore PA, Olsen HS, Migone TS, Sosnovtseva S, Carrell JA, Feng P, Giri JG et al (2001) Synthesis and release of B-lymphocyte stimulator from myeloid cells. *Blood* 97: 198–204

96   Scapini P, Nardelli B, Nadali G, Calzetti F, Pizzolo G, Montecucco C, Cassatella MA (2003) G-CSF-stimulated neutrophils are a prominent source of functional BLyS. *J Exp Med* 197: 297–302

97   He B, Raab-Traub N, Casali P, Cerutti A (2003) EBV-encoded latent membrane protein 1 cooperates with BAFF/BLyS and APRIL to induce T cell-independent Ig heavy chain class switching. *J Immunol* 171: 5215–5224

98   Fu L, Lin-Lee YC, Pham LV, Tamayo A, Yoshimura L, Ford RJ (2006) Constitutive NF-kappaB and NFAT activation leads to stimulation of the BLyS survival pathway in aggressive B-cell lymphomas. *Blood* 107: 4540–4548

99   Ohata J, Zvaifler NJ, Nishio M, Boyle DL, Kalled SL, Carson DA, Kipps TJ (2005) Fibroblast-like synoviocytes of mesenchymal origin express functional B cell-activating factor of the TNF family in response to proinflammatory cytokines. *J Immunol* 174: 864–870

100 Rolink AG, Tschopp J, Schneider P, Melchers F (2002) BAFF is a survival and maturation factor for mouse B cells. *Eur J Immunol* 32: 2004–2010

101 Do RK, Hatada E, Lee H, Tourigny MR, Hilbert D, Chen-Kiang S (2000) Attenuation of apoptosis underlies B lymphocyte stimulator enhancement of humoral immune response. *J Exp Med* 192: 953–964

102 Khare SD, Sarosi I, Xia XZ, McCabe S, Miner K, Solovyev I, Hawkins N, Kelley M, Chang D, Van G et al (2000) Severe B cell hyperplasia and autoimmune disease in TALL-1 transgenic mice. *Proc Natl Acad Sci USA* 97: 3370–3375

103 Mackay F, Woodcock SA, Lawton P, Ambrose C, Baetscher M, Schneider P, Tschopp J, Browning JL (1999) Mice transgenic for BAFF develop lymphocytic disorders along with autoimmune manifestations. *J Exp Med* 190: 1697–1710

104 Mackay F, Browning JL (2002) BAFF: A fundamental survival factor for B cells. *Nat Rev* 2: 465–475

105 Tangye SG, Bryant VL, Cuss AK, Good KL (2006) BAFF, APRIL and human B cell disorders. *Semin Immunol* 18: 305–317

106 Litinskiy MB, Nardelli B, Hilbert DM, He B, Schaffer A, Casali P, Cerutti A (2002) DCs induce CD40–independent immunoglobulin class switching through BLyS and APRIL. *Nat Immunol* 3: 822–829

107 Craxton A, Magaletti D, Ryan EJ, Clark EA (2003) Macrophage- and dendritic cell-dependent regulation of human B-cell proliferation requires the TNF family ligand BAFF. *Blood* 101: 4464–4471

108 Stein JV, Lopez-Fraga M, Elustondo FA, Carvalho-Pinto CE, Rodriguez D, Gomez-Caro R, De Jong J, Martinez AC, Medema JP, Hahne M (2002) APRIL modulates B and T cell immunity. *J Clin Invest* 109: 1587–1598

109 Roschke V, Sosnovtseva S, Ward CD, Hong JS, Smith R, Albert V, Stohl W, Baker KP, Ullrich S, Nardelli B et al (2002) BLyS and APRIL form biologically active heterotrimers

that are expressed in patients with systemic immune-based rheumatic diseases. *J Immunol* 169: 4314–4321

110 Avery DT, Kalled SL, Ellyard JI, Ambrose C, Bixler SA, Thien M, Brink R, Mackay F, Hodgkin PD, Tangye SG (2003) BAFF selectively enhances the survival of plasmablasts generated from human memory B cells. *J Clin Invest* 112: 286–297

111 Zhang X, Park CS, Yoon SO, Li L, Hsu YM, Ambrose C, Choi YS (2005) BAFF supports human B cell differentiation in the lymphoid follicles through distinct receptors. *Int Immunol* 17: 779–788

112 Cheema GS, Roschke V, Hilbert DM, Stohl W (2001) Elevated serum B lymphocyte stimulator levels in patients with systemic immune-based rheumatic diseases. *Arthritis Rheum* 44: 1313–1319

113 Seyler TM, Park YW, Takemura S, Bram RJ, Kurtin PJ, Goronzy JJ, Weyand CM (2005) BLyS and APRIL in rheumatoid arthritis. *J Clin Invest* 115: 3083–3092

114 Cohen SB, Emery P, Greenwald MW, Dougados M, Furie RA, Genovese MC, Keystone EC, Loveless JE, Burmester GR, Cravets MW et al (2006) Rituximab for rheumatoid arthritis refractory to anti-tumor necrosis factor therapy: Results of a multicenter, randomized, double-blind, placebo-controlled, phase III trial evaluating primary efficacy and safety at twenty-four weeks. *Arthritis Rheum* 54: 2793–2806

115 Edwards JC, Szczepanski L, Szechinski J, Filipowicz-Sosnowska A, Emery P, Close DR, Stevens RM, Shaw T (2004) Efficacy of B-cell-targeted therapy with rituximab in patients with rheumatoid arthritis. *N Engl J Med* 350: 2572–2581

116 Emery P, Fleischmann R, Filipowicz-Sosnowska A, Schechtman J, Szczepanski L, Kavanaugh A, Racewicz AJ, van Vollenhoven RF, Li NF, Agarwal S et al (2006) The efficacy and safety of rituximab in patients with active rheumatoid arthritis despite methotrexate treatment: Results of a phase IIB randomized, double-blind, placebo-controlled, dose-ranging trial. *Arthritis Rheum* 54: 1390–1400

117 Emery P BF, Martin-Mola, Pavelka L et al (2007) Relationship between peripheral B cell levels and loss of EULAR response in rheumatoid arthritis patients treated with rituximab. *Ann Rheum Dis* 66: 124

118 Dass S VE, Bingham SJ, Emery P (2006) The safety and efficacy of rituximab in patients with rheumatoid arthritis outside clinical trials: Real life experience. *Rheumatology* 45: i47

119 van Vollenhoven RF, Emery P, Bingham C, Keystone E, Greenwald M, Moreland LW, Kim D, Cooper S, Wagner B, Ward P (2006) Safety of rituximab in rheumatoid arthritis: Results of a pooled analysis. *Ann Rheum Dis* 65: 332

120 Tak PP CS, Emery P (2006) Baseline autoantibody status (RF, Anti-CCP) and clinical response following the first treatment course with rituximab. *Arthritis Rheum* 54: S368

121 Dass S RA, Vital EM, Jain S, Bingham SJ, McGonagle D, Emery P (2006) Highly sensitive B cell analysis predicts response to rituximab therapy in RA. *Arthritis Rheum* 54: S832

122 Emery P BF, Martin-Mola E, Pavelka K, Szczepanska L, Hagerty D, Margrini F, Beh-

rendt C, Kelman A (2007) Relationship between peripheral B cell levels and loss of EULAR response in rheumatoid arthritis patients treated with rituximab. *Ann Rheum Dis* 66: 124

123 Leandro MJ, Cambridge G, Ehrenstein MR, Edwards JC (2006) Reconstitution of peripheral blood B cells after depletion with rituximab in patients with rheumatoid arthritis. *Arthritis Rheum* 54: 613–620

124 Emery P FD, Ferracioli G, Udell J, van Vollenhoven RF, Rowe K, Agarwal S, Shaw T (2006) Long-term efficacy and safety of a repeat course of rituximab in RA patients with an inadequate response to disease modifying anti-rheumatic drugs. *Arthritis Rheum* 54: S228

125 Keystone EC FR, Emery P, Chubrick A, Dougados M, Baldassare AR, Bathon JM, Hessey E, Totoritis M, Cooper S (2006) Long-term efficacy and safety of a repeat treatment course of rituximab in rheumatoid arthritis patients with an inadequate response to one or more TNF inhibitors. *Arthritis Rheum* 54

126 Van Vollenhoven RF CS, Pavelka K, Kavanaugh A, Tak PP, Greenwald M, Cravets M, Ward P, Agarwal S, Magrini F (2006) Response to rituximab in patients with rheumatoid arthritis is maintained by repeat therapy: Results of an open-label trial. *Ann Rheum Dis* 65: 510

127 Emery P ST, Lehane PB et al (2004) Efficacy and safety of Rituximab at 2 years following a single treatment in patients with active rheumatoid arthritis. *Arthritis Rheum* 50: S659

128 Strand V B-GA, Pvelka K et al (2005) Two year improvements in physical function reflect sustained benefit in rheumatoid arthritis patients previously treated with rituximab. *Ann Rheum Dis* 64

129 Vugmeyster Y, Beyer J, Howell K, Combs D, Fielder P, Yang J, Qureshi F, Sandlund B, Kawaguchi L, Dummer W et al (2005) Depletion of B cells by a humanized anti-CD20 antibody PRO70769 in Macaca fascicularis. *J Immunother* 28: 212–219

130 Genovese MC KJ, Kohen MD, Lowenstein MB, Del Giudice J, Baldassare AR, Schechtman J, Gujrathi S, Trapp RG, Sweiss NJ, Spaniolo DG, Dummer W (2006) Safety and clinical activity of ocrelizumab (a humanised antibody targeting CD20+ B cells) in combination with methotrexate in moderate-severe rheumatoid arthritis patients (Ph I/II ACTION study). *Arthritis Rheum* 54: S66

131 Ostergaard M WC, Dawes PT, Rigby W, Petersen J, Kastberg H, Sierakowski S (2006) First clinical results of Humax-CD20 fully human monoclonal IgG1 antibody treatment in rheumatoid arthritis. Presented at the *Annual European Congress of Rheumatology*, abstract P0018

132 McKay J C-SH, Boling E, Valente R, Limanni A, Racewicz A, Wierbinska-Zarowny D, Fernandez V, Zhong J, Zilberstein M, Freimuth W (2005) Belimumab, a fully human monoclonal antibody to B-Lymphocyte stimulator, combined with standard of care therapy reduces the signs and symptoms of rheumatoid arthritis in a heterogeneous subject population. *Arthritis Rheum* 52: S710

133 Stohl W CW, Weisman M, Furie R, Weinstein A, Mishra N, Chevrier M, Fernandez V,

Migone TS, Freimuth W (2005) Belimumab, a novel fully human monoclonal antibody to B-Lymphocyte Stimulator, selectively modulates B cell subpopulations and immunoglobulins in a heterogeneous rheumatoid arthritis population. *Arthritis Rheum* 52: S444

134 Peano S PR, Bertolino M, Vigna E, Yu P, Visich J (2005) Nonclinical safety, pharmacokinetics and pharmacodynamics of TACI-Ig, a soluble receptor fusion protein antagonist of BLyS and APRIL. *Arthritis Rheum* 52: S285

135 Tak PP, Thurlings RM, Rossier C, Nestorov I, Dimic A, Mircetic V, Rischmueller M, Nasonov E, Shmidt E, Emery P, Munafo A (2008) Atacicept in patients with rheumatoid arthritis: Results of a multicenter, Phase Ib, double-blind, placebo-controlled, dose-escalating, single- and repeated-dose study. *Arthritis Rheum* 58: 61–72

136 Vugmeyster Y, Seshasayee D, Chang W, Storn A, Howell K, Sa S, Nelson T, Martin F, Grewal I, Gilkerson E et al (2006) A soluble BAFF antagonist, BR3-Fc, decreases peripheral blood B cells and lymphoid tissue marginal zone and follicular B cells in cynomolgus monkeys. *Am J Pathol* 168: 476–489

137 Fleischmann R WN, Shaw M, Birbara C, Anand B, Gujrathi S, Hendricks R, Rao T, Ren S, Weingart M, Wagner B, McLean L (2006) BR3-Fc phase I study: Safety, pharmacokinetics and pharmacodynamic effects of a novel BR3-Fc fusion protein in patients with rheumatoid arthritis. *Arthritis Rheum* 54: S229

Co-stimulatory pathways in the therapy of rheumatoid arthritis

Eric M. Ruderman and Richard M. Pope

Division of Rheumatology, Northwestern University Feinberg School of Medicine, McGaw M300, 240 E. Huron Street, Chicago, IL 60611, USA

Abstract

Although T lymphocytes are widely recognized as important effector cells in the immunopathogenesis of rheumatoid arthritis (RA), therapies targeting T cell populations have not been clinically successful, largely due to the toxicity associated with nonspecific T cell depletion. An alternative approach involves targeting T cell activation, a process that requires two distinct signals. In addition to the cognate interaction between the T cell receptor on T cells and antigen bound to the major histocompatibility complex on the antigen-presenting cell (APC), a second, co-stimulatory, signal is required for T cell activation. Therapies targeting co-stimulatory pathways, aimed at modifying the activation of T cells, rather than reducing their absolute numbers, may be an effective alternative to T cell depletion in RA and other rheumatic diseases. One such treatment, abatacept (CTLA4Ig), a fusion protein combining cytotoxic T lymphocyte antigen 4 (CTLA4) and a portion of the Fc domain of human IgG1, has been approved in the United States and the European Union for the treatment of RA. Abatacept may modulate the T cell or the APC to produce several different outcomes within the joint, including down-regulation of T cell activation, stimulation of T cell apoptosis, or possibly modulation of T regulatory cell activity. In large, controlled trials in patients with RA with an inadequate response to either methotrexate or TNF antagonists, abatacept effectively reduced disease activity. In the methotrexate-inadequate responder population, radiographic progression was slowed when compared to continued treatment with methotrexate alone. The safety profile of this therapy is similar to that of other biological response modifiers, with infection being the most concerning treatment-emergent adverse event. Additional co-stimulatory pathways may offer attractive targets in RA and other immune-mediated diseases, although, to date, none has had the clinical success of abatacept and the related CTLA4Ig fusion protein, belatacept.

Introduction

T cell-targeted therapy has been attempted in rheumatoid arthritis (RA) for many years. These efforts have included lymphatic duct drainage, total lymphoid irradiation, inhibition of T cell trafficking and T cell depletion with antibodies targeting T lymphocyte surface markers. While some of these therapies initially showed promising results, none turned out to be clinically effective and all have been hampered

New Therapeutic Targets in Rheumatoid Arthritis, edited by Paul-Peter Tak
© 2009 Birkhäuser Verlag Basel/Switzerland

by the toxicity associated with nonspecific T cell depletion. An alternate approach, the modulation of T cell activity by targeting co-stimulatory signaling, has recently been shown to be both effective and safe. This chapter describes the rationale for co-stimulatory therapy and reviews the data supporting its use in the treatment of RA.

The premise for this approach to therapy is based upon the fact that the activation of T cells requires two distinct signals. The first is cognate interaction between the T cell receptor and nominal antigen presented in the context of the major histocompatibility complex (MHC) on the surface of an antigen-presenting cell (APC). The second is provided by a number of receptors on the surface of T cells, which interact with ligands expressed on APCs. Co-stimulation is particularly important in the initial T cell response, promoting T cell activation and proliferation, so that therapies targeting this pathway have the potential to be very effective in diseases in which T cell activation and signaling play a critical role. Because this co-stimulatory signal is not unique to the antigen that triggers the primary T cell activation, it becomes feasible to target co-stimulatory pathways in a disease such as RA, where the identity of the inciting antigen remains unknown.

CD28-CD80/86 co-stimulation

One of the most prominent T cell co-stimulatory signals is mediated through the interaction of the cell surface protein CD28 on T cells and its ligands on APCs (reviewed in [1]). CD28 is constitutively expressed on most T cells, and it binds to both CD80 (B7-1) and CD86 (B7-2), which are expressed on APCs, including dendritic cells, B cells, and macrophages (Fig. 1A). CD80 and CD86 are increased during dendritic cell maturation, which enhances the ability of mature dendritic cells to activate T cells [2]. Macrophages from the joints of patients with RA express increased levels of CD80 and CD86 compared to RA peripheral blood monocytes [3]. CD80 and CD86 are not only expressed on APCs within the RA joint, they are also expressed on activated T cells, suggesting a potential self-sustaining mechanism for T cell activation [4]. Binding of CD28 to CD80 or CD86 provides the second signal required for maximal T cell activation; the absence of such a signal may result in anergy and apoptotic cell death.

Cytotoxic T lymphocyte antigen 4 (CTLA4 or CD152) is up-regulated following T cell activation, and it also interacts with CD80 and CD86, providing an important control for regulating T cell function [5, 6]. CTLA4 has a higher affinity for CD80 and CD86 and can displace CD28 from its interaction with these molecules, interrupting the activation signal [7] (Fig. 1B). Since the level of CTLA4 expression on activated T cells is proportional to the strength of T cell receptor signal, this pathway provides a mechanism to down-regulate T cell activation [8]. CTLA4 is

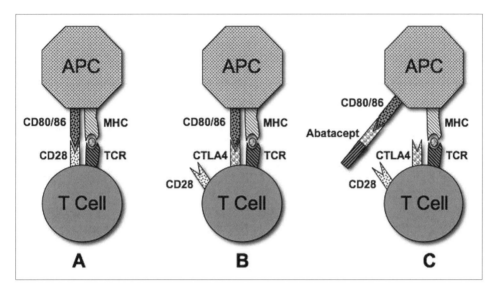

Figure 1
Abatacept (CTLA4Ig) interrupts co-stimulatory signals. T cell activation requires presenta-
tion of antigen presented by MHC molecules on antigen presenting cells (APC) together
with co-stimulation through CD28 and CD80/86 (A). Following T cell activation CTLA4 is
up-regulated and its affinity for CD80/86 is greater than that of CD28, which interrupts co-
stimulation, suppressing T cell activation (B). CTLA4Ig is capable of interrupting signaling
through CD28 and potentially through CTLA4 (C).

expressed on T cells from the RA joint [4], supporting the potential importance of
this pathway in regulating T cell activation in this disease.

In addition to interfering with activation, CTLA4-CD80/86 interaction may
also provide negative signals that lead to long-term tolerance. The binding of cell
surface CTLA4 (or with CTLA4Ig) to CD80/86 on dendritic cells may induce the
production of indoleamine 2,3-dioxygenase (IDO) by the dendritic cell [9]. IDO is
an enzyme that promotes local tryptophan depletion, resulting in the inhibition of
T cell proliferation and the induction of apoptosis [10].

CTLA4 signaling may also play a role in the development of an important class
of T cells, CD4$^+$CD25$^+$ T regulatory cells. The development of T regulatory cells
is dependent upon the transcription factor forkhead-box protein 3 (FoxP3), since
deficiency of FoxP3 results in the lack of T regulatory cells and autoimmune disease
[11]. Interactions between CD28 and CD80/86 are necessary for the generation of
T regulatory cells [12]. Increased numbers of CD4$^+$CD25$^+$ T regulatory cells are
present in the RA joint, and they are highly active at suppressing the activation
of other T cells *ex vivo* [13]. However, in the presence of tumor necrosis factor

(TNF)-α, which is present in the RA joint, the local T regulatory cells demonstrate a diminished capacity to suppress T cell activation [3]. Together, these observations support the potential role of CTLA4 expressed in the RA joint in the development of T regulatory cells.

Other co-stimulatory pathways

Inducible co-stimulatory molecule (ICOS) and programmed death-1 (PD-1) are CD28 family members also expressed on T cells following activation by T cell receptor ligation and CD28-CD80/86 co-stimulation [1]. The ligands for ICOS, B7-related protein 1 (B7RP.1 or B7h), and for PD-1 (PD-L1 and PD-L2) are expressed on APCs [1]. Co-stimulation through CD28 and ICOS are different, since interfering with CD28 during the induction of the immune response suppresses inflammation, while interfering with ICOS during the effector phase reduces inflammation [14]. Both ICOS and B7RP.1 are highly expressed on RA synovial T cells, while B7RP.1 is strongly expressed on synovial macrophages, suggesting an important role in RA [15]. Supporting a potential role in RA, ICOS was essential for developing collagen-induced arthritis and blocking ICOS-B7 interactions suppressed collagen-induced arthritis [16, 17].

PD-1 acts as a negative regulator to T cell activation, and ligation by PD-L1 or PD-L2 results in decreased T cell proliferation and cytokine production [1]. PD-1 is overexpressed on RA synovial T cells and PD-LI on synovial macrophages [18]. However, a soluble form of PD-1 detected in RA synovial fluid was capable of over-riding the suppressive effect of cell surface PD-1 [18]. These observations support an important role for PD-1 in RA and identify it as a potential target for the development of future therapeutic intervention.

Another co-stimulatory pathway that may be relevant to RA is the CD40-CD154 (CD40L) pathway. CD40L is up-regulated on activated T cells, whereas CD40 is expressed on APCs [19]. CD40L is increased on RA synovial fluid T cells and CD40 is expressed on RA synovial fibroblasts, and interruption of these interactions with an antibody to CD40L suppressed fibroblast activation [20]. While these observations suggest that this pathway may be an attractive therapeutic target in RA, studies described later in this chapter indicate that the use of an antibody to CD40L may be too toxic.

Pre-clinical development of CTLA4Ig

Two recombinant molecules (CTLA4Ig) have recently been developed that combine the extracellular domain of human CTLA4 with a portion of the Fc domain of IgG1. One of these fusion proteins, abatacept, binds CD80 more avidly than

CD86. A second-generation version of this molecule, belatacept, with two amino acid mutations, has increased binding avidity for CD86 [21]. CTLA4Ig also interrupts signaling through the interaction of CD80/86 and cell surface CTLA4, which may potentially affect the development or function of T regulatory cells, and interrupt antigen-specific tolerance (Fig. 1C). On the other hand, inhibition of signaling through CTLA4 has been shown to promote Th2 development, which may be beneficial in RA [4].

The alternate approach to T cell therapy, limiting activation by blocking co-stimulatory signals, has proven effective in clinical trials, and the first agent in this class, abatacept, is now approved for clinical use in the U.S and Europe. By interrupting one of the more prominent and well-characterized T cell co-stimulatory pathways, abatacept might be expected to affect the development of T cell-mediated synovitis. Early work in an animal model of inflammatory arthritis, collagen-induced arthritis in the BB rat, showed that pretreatment with CTLA4Ig was able to prevent the development of arthritis [22]. Additional animal data have shown that CD28-deficient mice are resistant to the development of collagen-induced arthritis, further highlighting the important role of the CD28-CD80/86 co-stimulatory pathway in the development and maintenance of inflammatory synovitis [23].

CTLA4Ig pilot clinical trials in RA

The first large clinical trial to examine the efficacy of co-stimulatory blockade in RA was a placebo-controlled, dose-ranging study of abatacept and belatacept, a second-generation modification of abatacept, in patients with refractory RA [21] (Tab. 1). Two hundred and fourteen patients, all of whom had failed treatment with at least one disease-modifying anti-rheumatic drug (DMARD), were randomized to receive four infusions of placebo, abatacept, or belatacept over 2 months. Prior therapy with etanercept, the only biological available at the time of the trial, was allowed, although the published manuscript does not state how many had actually received this treatment. Three doses of each of the two compounds were tested (0.5, 2, or 10 mg/kg), and these were given at baseline, 2, 4, and 8 weeks. This dosing strategy was selected to provide a loading dose that could lead to a more rapid response, to be followed by less frequent maintenance dosing. Abatacept was given as monotherapy in this trial; all background DMARD therapy was stopped. Both compounds proved effective at generating the primary endpoint, an ACR20 response, and both had a clear dose response. Abatacept appeared somewhat more effective at producing higher levels of response (ACR50 and ACR70), although the trial was not powered to show statistical differences between the two compounds. No further RA trials have studied belatacept, and this compound is currently being pursued for transplant indications, as discussed below [24]. No significant safety signals were seen in this trial, even at the higher dose levels.

Table 1 - Abatacept efficacy in blinded, controlled clinical trials.[a]

Trial	n	Trial duration	Study population	Background therapy	ACR20 on abatacept	ACR20 on placebo	Remission on abatacept (DAS28 <2.6)
Moreland et al. [21]	122	12 weeks	Active RA	None	53%	31%	N/A
Kremer et al. [27]	339	6 months	Methotrexate inadequate responders	Methotrexate	60.0%	35.3%	26.1%
Kremer et al. [25]	339	12 months	Methotrexate inadequate responders	Methotrexate	62.6%	36.1%	34.8%
Kremer et al. [26]	652	1 year	Methotrexate inadequate responders	Methotrexate	73.1%	39.7%	23.8%
Schiff et al. [38]	432	6 months	Methotrexate inadequate responders	Methotrexate	66.7% (59.4% infliximab)	41.8%	N/A
Weinblatt et al. [34]	121	6 months	Etanercept inadequate responders	Etanercept	48%	28%	N/A
Genovese et al. [32]	391	24 weeks	Anti-TNF inadequate responders	Methotrexate	50.4%	19.5%	10%
Giannini et al. [37]	122	6 months	Active JIA	None specified (77% taking methotrexate)	20% flared	53.2% flared	N/A

[a]All trials dosed abatacept at or approximating 10 mg/kg, except Weinblatt et al., which dosed abatacept at 2 mg/kg. Schiff et al. had an additional arm that was treated with infliximab 3 mg/kg. Giannini et al. used an endpoint of disease flare on blinded therapy, following an open-label run-in to define abatacept responders.

Following the dose-ranging trial, abatacept has been studied in two published trials in RA patients showing an inadequate response to methotrexate [25–27] (Tab. 1). Both trials had an initial placebo-controlled period, followed by an open label extension treatment period. Both used the dosing schedule established in the pilot study: three loading doses given 2 weeks apart, followed by dosing every 4 weeks thereafter. The first (Phase II) study included two doses, 2 and 10 mg/kg, compared with placebo, with blinded treatment continued for 6 months [25, 27]. The second, pivotal trial compared placebo with a fixed dose that was determined by patient weight range, approximating 10 mg/kg (i.e., 500 mg for patients under 60 kg, 750 mg for those 60–100 kg, and 1000 mg for those over 100 kg), with blinded therapy continued for a year [26]. This latter dosing strategy has become the commercially approved dose.

CTLA4Ig Phase II trials

In the Phase II trial, 339 patients were randomized to receive abatacept 2 or 10 mg/kg vs placebo [27] (Tab. 1). The population studied had long-standing RA (approximately 9 years) and had failed multiple prior DMARDs. Unlike the pilot study, methotrexate was continued during the trial, along with abatacept or placebo infusions. The mean duration of methotrexate therapy was just under 3 years, and the mean dose was just over 15 mg/week. Abatacept at 10 mg/kg was clinically effective, with 60% achieving the primary ACR20 endpoint at 6 months, compared with 35% of the placebo-treated patients, a difference that was highly statistically significant. This dose of abatacept also achieved higher levels of response, with 37% and 17% achieving ACR50 and ACR70 responses, respectively, compared with 12% and 2% of the placebo-treated patients. The lower dose of abatacept (2 mg/kg) was not as effective; it produced statistically better responses than placebo for ACR50 and ACR70, but not for ACR20, the primary endpoint.

Blinded therapy in the Phase II trial with abatacept was continued for 12 months, and the 12-month data have been published separately [25] (Tab. 1). At 12 months, the 10 mg/kg dose continued to show a marked improvement in ACR20 response relative to the placebo group (63% vs 36%). ACR50 and ACR70 responses also continued to show benefit relative to placebo at this dose (42% vs 20% and 21% vs 8%, respectively). Remission rates also favored abatacept; 35% of the 10 mg/kg treatment group met DAS28 criteria for remission at 12 months, compared with 10% of the placebo group. Adverse events, including serious adverse events, were similar between the abatacept and placebo groups at both 6 and 12 months [25, 27]. In particular, there was no evidence of an increased risk of infection, and no opportunistic infections were reported in the abatacept-treated group. There were three malignancies reported in the placebo group and four in the abatacept groups (all at 10 mg/kg); none of the latter was felt by investigators to be related to the study drug. It should be noted that subjects in this trial were permitted to adjust their methotrexate and steroid doses, or to add additional non-biological DMARDs, during the second 6 months of therapy; the number who actually did so was not reported.

Twelve months of therapy with abatacept in the phase II trial produced significant improvements in function and health-related quality of life (HRQOL) measurements relative to placebo [26, 28]. The health assessment questionnaire (HAQ), which showed only modest functional impairment at baseline (mean score 1.0), was nevertheless significantly improved in the 10 mg/kg group relative to placebo (42% vs 10%). Compared with placebo, there were statistically significant improvements in all eight scales of the Short Form-36 (SF-36) as well as the summary scores of the physical component and mental component. Improvements were greater than one-half standard deviation, which has been described as clinically significant [29], and were of a magnitude similar to that seen with other effective disease modifying therapies in RA. The greatest effect sizes were seen for the physical function, bodily pain, and vitality scales. Improvement in HRQOL was only marginal for the 2 mg/kg dose and was related to the clinical response for the 10 mg/kg dose.

Perhaps not surprisingly, biomarkers of inflammation also showed improvement with abatacept therapy in the Phase II study [30]. Interleukin (IL)-6, soluble IL-2 receptor (sIL-2R), C-reactive protein (CRP), soluble E-selectin, and soluble intercellular adhesion molecule-1 were all significantly reduced in the 10 mg/kg treatment group compared with placebo. The only one of these molecules produced by the activated T cells directly impacted by abatacept is sIL2-R, so the implication of this observation is that abatacept has downstream effects on other immunologically active cells. No radiographic data were collected in this trial.

CTLA4Ig Phase III

In the AIM (abetacept in methotrexate inadequate responders) trial, the largest placebo-controlled trial to date in methotrexate inadequate responders, 652 patients were randomized 2:1 to treatment with either abatacept or placebo at a fixed dose approximating 10 mg/kg [26] (Tab. 1). These patients also had long-standing disease (mean 8.6 years); according to the study criteria, all were taking methotrexate at enrollment (mean dose just under 16 mg/week), and most (70.9%) were receiving low-dose corticosteroids. Response at 6 months was similar to that seen in the Phase II trial, with 68% of the abatacept-treated patients achieving an ACR20, compared with 40% of the placebo-treated patients. Efficacy was also demonstrated at higher levels of response, with 40% of abatacept-treated patients achieving an ACR50 and 20% achieving an ACR70, compared with 17% and 7% in the placebo group.

During the second 6 months of blinded therapy in this study, there was increasing benefit associated with abatacept treatment. ACR20/50/70 responses improved to

73%, 48%, and 29%, compared with 40%, 18%, and 6% for placebo; for ACR50 and ACR70, the increased proportion of responders was statistically significant. In addition, the percentage of abatacept-treated patients achieving remission by DAS28 criteria increased from 15% at 6 months to 24% at 1 year. As in the Phase II trial, methotrexate dose adjustments and the addition of a second non-biological DMARD were allowed during the second 6 months of the study, although subjects and investigators remained blinded to treatment assignment. Additional therapy was begun in just 8% of the abatacept-treated patients compared with 14% of the placebo group, suggesting that the increasing likelihood of higher level response seen with longer-term therapy was not attributable to modification of medications other than abatacept.

Both function and HRQOL were improved by abatacept in this study [26, 31]. Baseline function was much more impaired than in the group studied in the Phase II trial, with a baseline HAQ disability index of 1.7 for both treatment groups. The mean improvement in HAQ score, as well as the proportion of patients with a clinically meaningful improvement, was statistically greater for the abatacept group at both 6 and 12 months. Significant improvements were also seen in both the physical and mental component summaries of the SF-36 at both time points.

The AIM study was the first to demonstrate structural benefit with abatacept. At 1 year, radiographic progression, measured using the Genant modification of the Sharp scoring system, was reduced by approximately 50% in the abatacept-treated group compared with placebo [26]. Although the radiographic data set was not complete (only 92% of study subjects had both baseline and at least one post-treatment radiograph), sensitivity analysis suggested that this effect was real.

Overall, adverse events were generally similar for abatacept and placebo in this study. The two most commonly reported adverse events, headache and nasopharyngitis, were more frequent in the abatacept group, and there were also more discontinuations due to adverse events in this group (4.2% vs 1.8%). Serious infections, in particular pneumonia, were numerically more common with abatacept therapy. Six abatacept-treated patients (1.4%) were classified as having pneumonia or bronchopneumonia, compared with a single placebo patient (0.5%). One case of tuberculosis was reported in each group, although neither was confirmed by culture.

CTLA4Ig in those with insufficient responses to TNF inhibitors

The effectiveness of abatacept has been confirmed in a second Phase III study in a different population. It is now well-recognized that TNF-α antagonists, while very effective for the signs and symptoms of RA, fail to achieve an adequate response, fail to maintain that response, or are poorly tolerated in a subset of patients. Abatacept has been evaluated in this population, where other approaches to controlling disease are obviously required. In the ATTAIN study, a group of 391 patients with long-standing active disease (mean just under 12 years) who had had an inadequate response or were intolerant to a TNF antagonist received blinded treatment with abatacept or placebo while continuing background DMARDs other than the TNF antagonist [32] (Tab. 1). The same fixed dose, approximating 10 mg/kg, was used, along with the standard dosing schedule.

In this trial, 50% of the abatacept-treated patients achieved an ACR20, the primary endpoint, compared with 20% of the placebo-treated patients. Abatacept-treated patients also did significantly better at higher levels of response, including ACR50 (20% vs 4%), ACR70 (4% vs 2%), and remission by DAS28 criteria (10% vs 1%). Patient-centered outcomes also improved in this study. Statistically significant improvements relative to placebo were seen in HAQ scores, fatigue, and both the physical and mental components of scores of the SF-36 [33]. Differences were generally seen within 2–3 months; in the case of fatigue, a difference could be seen after just 4 weeks of therapy (two doses of abatacept). These were notable improvements in patients who had previously had an inadequate response to TNF antagonists, the best previously available therapy. No additional safety signals were seen in the ATTAIN trial beyond those previously seen in the trials with methotrexate inadequate responders [32].

A second trial also looked at abatacept treatment in patients failing a TNF antagonist, although it was designed to evaluate combination therapy, rather than a switch to abatacept, in the hopes that targeting two distinct pathways in the pathophysiology of RA would have a greater effect [34] (Tab. 1). In this trial, 121 patients with persistently active disease despite receiving etanercept 25 mg twice weekly (the standard dose at the time the study was undertaken) were randomized 2:1 to receive abatacept 2 mg/kg or placebo, in addition to etanercept, for 1 year. For the primary endpoint in this study, the proportion of ACR20 responders at 6 months, there was no demonstrable benefit with abatacept (48% vs 30%, difference not statistically significant). ACR50 responses at 6 months also did not differ, although there was a statistically significant difference in ACR70 responses (11% vs 0%, $p = 0.042$). Safety, however, was compromised by combination therapy in this trial. The incidence of overall serious adverse events (16.5% vs 2.8%), serious adverse events judged related to study drug (5.9% vs 0%), and serious infections (3.5% vs 0%) during 1 year of blinded therapy were all greater with the combination of abatacept and etanercept.

Ultimately, 2 mg/kg, as discussed above, was found to be a sub-therapeutic dose of abatacept. Subjects in this combination trial were allowed to enroll into an open-label extension study, during which they received abatacept at 10 mg/kg. The pattern of increased adverse events seen in the 80 subjects who entered this 2-year extension was similar, with 32.5% reporting serious adverse events, 3.8% reporting serious adverse events judged related to therapy, and 1 patient (1.3%) developing a serious infection (septic arthritis). Three patients developed malignancies during this extension phase.

An additional, larger trial has expanded the evaluation of the safety of abatacept in combination with other background therapies [35]. In this 1-year, blinded trial, patients with active disease despite treatment with at least one biological or non-biological DMARD, were randomized 2:1 to receive abatacept at the fixed dose approximating 10 mg/kg or placebo, in addition to their existing therapy. While the clinical outcomes captured in this study were limited, it did confirm that the addition of abatacept leads to significant improvements in patient pain assessments, patient and physician global assessments, and HAQ scores over 1 year.

With respect to safety outcomes, the primary goal of this study, there were important signals seen with the addition of abatacept to background therapy. Overall incidence of adverse events, serious adverse events, and discontinuations due to adverse events did not show any differences between the groups receiving abatacept or placebo. Serious infections, however, were numerically more frequent in the abatacept-treated group (2.9% vs 1.9%). Perhaps more importantly, in the subset of patients receiving therapy with a background biological DMARD (primarily TNF antagonists), the incidence of serious adverse events (22.3% vs 12.5%) and serious infections (5.8% vs 1.6%) was markedly elevated in the abatacept-treated group, confirming the additional risk seen with combination biological therapy seen in the earlier study.

Overall safety of CTLA4Ig

An analysis of the integrated safety findings of five clinical trials of abatacept in RA evaluated 1955 patients treated with abatacept and 989 patients treated with placebo [36]. While the incidence of adverse events was generally similar between the two groups, study discontinuations due to adverse events and serious adverse events were slightly higher for the abatacept groups (5.5% vs 3.9% and 2.7% vs 1.6%). The overall incidence of serious infection was 3.0% with abatacept and 1.9% with placebo; reported malignancies were 1.2% with abatacept and 1.0% with placebo. The most commonly reported adverse events with abatacept were headache, upper respiratory infection, nausea, and nasopharyngitis.

Juvenile inflammatory arthritis

Two additional controlled trials of abatacept, both recently published, are worth noting. In a trial of abatacept for juvenile inflammatory arthritis, 122 of 190 subjects met criteria for improvement after 4 months of open-label therapy with abatacept and were randomized to receive blinded therapy with abatacept or placebo for an additional 6 months [37] (Tab. 1). During the blinded portion of the trial, 53% of the placebo-treated patients flared, compared with just 20% of those treated with abatacept ($p = 0.0003$). No patients in either group withdrew from the study due to an adverse event. The overall incidence of adverse events was similar during abatacept and placebo treatment, and there were no serious adverse events in the abatacept group during the double-blind period.

Direct comparison of CTLA4Ig with anti-TNFα

In another trial, the first blinded comparison of two biological therapies in 431 patients with RA, treatment with abatacept ~10 mg/kg every 4 weeks was compared with treatment with infliximab 3 mg/kg every 8 weeks over 1 year [38]. During the first 6 months, a third group received placebo infusions every 4 weeks; these subjects were switched to abatacept therapy during the last 6 months of the trial. There was no statistical difference between abatacept and infliximab for the primary endpoint of reduction in DAS28 at 6 months; both groups showed statistically greater clinical improvement than placebo. At 1 year the improvement of the DAS28 in the abatacept-treated patients (2.88 DAS28 units) was modestly, but statistically, greater than the improvement in the infliximab group (2.25 DAS28 units). Safety outcomes in this study favored abatacept, with 18.2% of the infliximab-treated subjects reporting serious adverse events over 1 year of double blind therapy, compared with 9.6% of the abatacept-treated subjects. Serious infections were reported in 8.5% of infliximab treated vs 1.9% of abatacept-treated patients during this same time period. The two reported cases of tuberculosis both occurred in infliximab-treated patients.

Abatacept mechanism of action in RA

One of the more interesting findings of the trials of abatacept in the methotrexate and TNF inadequate responders is that the drug worked so well in patients with many years of disease. From its purported mechanism, to affect the induction of the immune response, one might assume that this treatment would be most effective early in the disease course, before the arthritogenic T cell clones have become fully activated. The demonstration of abatacept's effectiveness later in the disease course has a number of potential explanations. The most obvious is that T cell activation is an ongoing process in RA, either through the cumulative addition of new populations of activated cells or the continued stimulation of the existing cells, and that abatacept is able to interrupt this ongoing process. Another possibility is that the clinical effect of abatacept relies on mechanisms other than simply interrupting the CD28-CD80/86 axis, such as inducing production of IDO *via* its ligation to APCs, which in turn may promote T cell tolerance. It is also possible that CD28 promotes activation of APCs, which is interrupted by CTLA4Ig, or that CTLA4Ig signaling through CD80/86 might promote the expression of suppressive cytokines such as IL-10. The generation and maintenance of T regulatory cells is mediated through CTLA4. The effect of abatacept on T regulatory cells in RA is unclear. However, treatment of patients with infliximab results in enhanced T regulatory cell function [39, 40], consistent with the recently described suppressive effects of TNF-α [3]. Therefore an understanding of the role of abatacept on T regulatory cells may provide insights into its mechanism of action.

Since T cell activation in RA likely starts even before clinical symptoms are apparent, it is possible that abatacept given earlier in the disease process may be even more effective. There are no data yet to support this theory, but trials are underway in early disease, including one study in patients first presenting with inflammatory arthritis that aims to determine whether modulation of co-stimulatory signals at this point may modify the course of disease, potentially preventing the development of full-blown RA (clintrials.gov). Abatacept is also being studied in systemic lupus erythematosus (SLE), although preliminary data did not demonstrate efficacy [41].

Belatacept

Belatacept, the modified version of abatacept used in several arms of the original dose-ranging study, has not been studied further in RA. This compound is currently being developed as an immunomodulatory treatment in organ transplantation. Belatacept compared favorably with cyclosporine at preventing acute rejection after renal transplantation [24]. Subjects treated with belatacept in this trial had higher mean glomerular filtration rates and a lower incidence of chronic allograft nephropathy at 12 months than those treated with cyclosporine.

Failed efforts at interfering with co-stimulatory pathways

While the CD80/86-CD28 interaction is the most well-characterized co-stimulatory signal involved in T cell activation, other co-stimulatory molecules may prove to be useful targets in the treatment of RA or other autoimmune diseases. The binding of CD40 to its ligand, CD40L/CD154, enhances T cell activation and stimulates B cell differentiation and proliferation [42]. This particular pathway may play a role in the maintenance of autoimmunity [43]. Two distinct antibodies against CD40L have been studied in human trials of SLE. One proved to be ineffective but well tolerated; the second demonstrated preliminary evidence of benefit in lupus nephritis but was associated with unacceptable thromboembolic toxicity [44, 45].

Another approach that failed was the use of a stimulatory anti-CD28 monoclonal antibody. A superagonistic anti-rat CD28 monoclonal antibody was effective at treating adjuvant-induced arthritis [46]. However, when a humanized superagonist antibody was administered to human volunteers, cytokine storm was induced resulting from the rapid induction of cytokines [47]. The patients developed pulmonary infiltrates, acute renal failure and intravascular coagulation which was life threatening [47]. Therefore, not all mechanisms of interfering with the CD28-CD80/86 pathway are safe.

Modulation of co-stimulatory pathways represents a novel approach to the treatment of immune-mediated diseases that seeks to normalize host immune responses, rather than to disrupt intercellular signaling or to reduce specific populations of immunoactive cells. Interruption of signaling at the CD80/86-CD28 axis has proven to be an effective therapy for RA; disruption of the CD40-CD40L axis has been less successful in SLE. Future approaches to this type of modulation may involve the targeting of additional co-stimulatory pathways, interference with the expression of co-stimulatory molecules, or disruption of the intracellular signaling pathways triggered by cell-surface molecule interaction.

References

1   Keir ME, Sharpe AH (2005) The B7/CD28 costimulatory family in autoimmunity. *Immunol Rev* 204: 128–43

2   De Smedt T, Pajak B, Muraille E, Lespagnard L, Heinen E, De Baetselier P et al (1996) Regulation of dendritic cell numbers and maturation by lipopolysaccharide *in vivo*. *J Exp Med* 184: 1413–24

3   van Amelsfort JM, van Roon JA, Noordegraaf M, Jacobs KM, Bijlsma JW, Lafeber FP et al (2007) Proinflammatory mediator-induced reversal of CD4+,CD25+ regulatory T cell-mediated suppression in rheumatoid arthritis. *Arthritis Rheum* 56: 732–42

4   Verwilghen J, Lovis R, De Boer M, Linsley PS, Haines GK, Koch AE et al (1994) Expression of functional B7 and CTLA4 on rheumatoid synovial T cells. *J Immunol* 153: 1378–85

5   Karandikar NJ, Vanderlugt CL, Walunas TL, Miller SD, Bluestone JA (1996) CTLA-4: A negative regulator of autoimmune disease. *J Exp Med* 184: 783–8

6   Walunas TL, Bakker CY, Bluestone JA (1996) CTLA-4 ligation blocks CD28-dependent T cell activation. *J Exp Med* 183: 2541–50

7   Peach RJ, Bajorath J, Brady W, Leytze G, Greene J, Naemura J et al (1994) Complementarity determining region 1 (CDR1)- and CDR3-analogous regions in CTLA-4 and CD28 determine the binding to B7-1. *J Exp Med* 180: 2049–58

8   Egen JG, Allison JP (2002) Cytotoxic T lymphocyte antigen-4 accumulation in the immunological synapse is regulated by TCR signal strength. *Immunity* 16: 23–35

9   Grohmann U, Orabona C, Fallarino F, Vacca C, Calcinaro F, Falorni A et al (2002) CTLA-4-Ig regulates tryptophan catabolism *in vivo*. *Nat Immunol* 3: 1097–101

10 Lee GK, Park HJ, Macleod M, Chandler P, Munn DH, Mellor AL (2002) Tryptophan

deprivation sensitizes activated T cells to apoptosis prior to cell division. *Immunology* 107: 452–60

11   Fontenot JD, Gavin MA, Rudensky AY (2003) Foxp3 programs the development and function of CD4⁺CD25⁺ regulatory T cells. *Nat Immunol* 4: 330–6

12   Tang Q, Henriksen KJ, Boden EK, Tooley AJ, Ye J, Subudhi SK et al (2003) Cutting edge: CD28 controls peripheral homeostasis of CD4⁺CD25⁺ regulatory T cells. *J Immunol* 171: 3348–52

13   van Amelsfort JM, Jacobs KM, Bijlsma JW, Lafeber FP, Taams LS (2004) CD4(+) CD25(+) regulatory T cells in rheumatoid arthritis: Differences in the presence, phenotype, and function between peripheral blood and synovial fluid. *Arthritis Rheum* 50: 2775–85

14   Liang L, Sha WC (2002) The right place at the right time: Novel B7 family members regulate effector T cell responses. *Curr Opin Immunol* 14: 384–90

15   Ruth JH, Rottman JB, Kingsbury GA, Coyle AJ, Haines GK 3rd, Pope RM et al (2007) ICOS and B7 costimulatory molecule expression identifies activated cellular subsets in rheumatoid arthritis. *Cytometry A* 71: 317–26

16   Nurieva RI, Treuting P, Duong J, Flavell RA, Dong C (2003) Inducible costimulator is essential for collagen-induced arthritis. *J Clin Invest* 111: 701–6

17   Iwai H, Kozono Y, Hirose S, Akiba H, Yagita H, Okumura K et al (2002) Amelioration of collagen-induced arthritis by blockade of inducible costimulator-B7 homologous protein costimulation. *J Immunol* 169: 4332–9

18   Wan B, Nie H, Liu A, Feng G, He D, Xu R et al (2006) Aberrant regulation of synovial T cell activation by soluble costimulatory molecules in rheumatoid arthritis. *J Immunol* 177: 8844–50

19   Liu EH, Siegel RM, Harlan DM, O'Shea JJ (2007) T cell-directed therapies: Lessons learned and future prospects. *Nat Immunol* 8: 25–30

20   Liu MF, Chao SC, Wang CR, Lei HY (2001) Expression of CD40 and CD40 ligand among cell populations within rheumatoid synovial compartment. *Autoimmunity* 34: 107–13

21   Moreland LW, Alten R, Van den Bosch F, Appelboom T, Leon M, Emery P et al (2002) Costimulatory blockade in patients with rheumatoid arthritis: A pilot, dose-finding, double-blind, placebo-controlled clinical trial evaluating CTLA-4Ig and LEA29Y eighty-five days after the first infusion. *Arthritis Rheum* 46: 1470–9

22   Knoerzer DB, Karr RW, Schwartz BD, Mengle-Gaw LJ (1995) Collagen-induced arthritis in the BB rat. Prevention of disease by treatment with CTLA-4-Ig. *J Clin Invest* 96: 987–93

23   Tada Y, Nagasawa K, Ho A, Morito F, Ushiyama O, Suzuki N et al (1999) CD28-deficient mice are highly resistant to collagen-induced arthritis. *J Immunol* 162: 203–8

24   Vincenti F, Larsen C, Durrbach A, Wekerle T, Nashan B, Blancho G et al (2005) Costimulation blockade with belatacept in renal transplantation. *N Engl J Med* 353: 770–81

25   Kremer JM, Dougados M, Emery P, Durez P, Sibilia J, Shergy W et al (2005) Treatment of rheumatoid arthritis with the selective costimulation modulator abatacept: Twelve-

month results of a Phase IIb, double-blind, randomized, placebo-controlled trial. *Arthritis Rheum* 52: 2263–71

26 Kremer JM, Genant HK, Moreland LW, Russell AS, Emery P, Abud-Mendoza C et al (2006) Effects of abatacept in patients with methotrexate-resistant active rheumatoid arthritis: A randomized trial. *Ann Inter Med* 144: 865–76

27 Kremer JM, Westhovens R, Leon M, Di Giorgio E, Alten R, Steinfeld S et al (2003) Treatment of rheumatoid arthritis by selective inhibition of T-cell activation with fusion protein CTLA4Ig. *N Engl J Med* 349: 1907–15

28 Emery P, Kosinski M, Li T, Martin M, Williams GR, Becker JC et al (2006) Treatment of rheumatoid arthritis patients with abatacept and methotrexate significantly improved health-related quality of life. *J Rheumatol* 33: 681–9

29 Norman GR, Sloan JA, Wyrwich KW (2003) Interpretation of changes in health-related quality of life: The remarkable universality of half a standard deviation. *Med Care* 41: 582–92

30 Weisman MH, Durez P, Hallegua D, Aranda R, Becker JC, Nuamah I et al (2006) Reduction of inflammatory biomarker response by abatacept in treatment of rheumatoid arthritis. *J Rheumatol* 33: 2162–6

31 Russell AS, Wallenstein GV, Li T, Martin MC, Maclean R, Blaisdell B et al (2007) Abatacept improves both the physical and mental health of patients with rheumatoid arthritis who have inadequate response to methotrexate treatment. *Ann Rheum Dis* 66: 189–94

32 Genovese MC, Becker JC, Schiff M, Luggen M, Sherrer Y, Kremer J et al (2005) Abatacept for rheumatoid arthritis refractory to tumor necrosis factor alpha inhibition. *N Engl J Med* 353: 1114–23

33 Westhovens R, Cole JC, Li T, Martin M, Maclean R, Lin P et al (2006) Improved health-related quality of life for rheumatoid arthritis patients treated with abatacept who have inadequate response to anti-TNF therapy in a double-blind, placebo-controlled, multi-centre randomized clinical trial. *Rheumatology* 45: 1238–46

34 Weinblatt M, Schiff M, Goldman A, Kremer J, Luggen M, Li T et al (2007) Selective costimulation modulation using abatacept in patients with active rheumatoid arthritis while receiving etanercept: A randomised clinical trial. *Ann Rheum Dis* 66: 228–34

35 Weinblatt M, Combe B, Covucci A, Aranda R, Becker JC, Keystone E (2006) Safety of the selective costimulation modulator abatacept in rheumatoid arthritis patients receiving background biologic and nonbiologic disease-modifying antirheumatic drugs: A one-year randomized, placebo-controlled study. *Arthritis Rheum* 54: 2807–16

36 Moreland L, Kaine J, Espinoza L, McCann T, Aranda R, Becker JC et al (2005) Safety of abatacept in rheumatoid arthritis patients in five double-blind, placebo-controlled trials. American College of Rheumatology Annual Meeting, Abstract 886

37 Ruperto N, Lovell DJ, Quartier P, Paz E, Rubio-Pérez NE, Silva CA et al (2008) Abatacept in children with juvenile idiopathic arthritis: a randomised, double-blind, placebo-controlled withdrawal trial. *Lancet* 372: 383–391

38 Schiff M, Keiserman M, Codding C, Songcharoen S, Berman A, Nayiager S et al (2008)

Efficacy and safety of abatacept or infliximab vs placebo in ATTET: a phase III, multi-centre, randomised, double-blind, placebo-controlled study in patients with rheumatoid arthritis and an inadequate response to methotrexate. *Ann Rheum Dis* 67: 1096–1103

39   Nadkarni S, Mauri C, Ehrenstein MR (2007) Anti-TNF-alpha therapy induces a distinct regulatory T cell population in patients with rheumatoid arthritis *via* TGF-beta. *J Exp Med* 204: 33–9

40   Ehrenstein MR, Evans JG, Singh A, Moore S, Warnes G, Isenberg DA et al (2004) Compromised function of regulatory T cells in rheumatoid arthritis and reversal by anti-TNFalpha therapy. *J Exp Med* 200: 277–85

41   Merill JT, Burgos-Vargas R, Westhovens R, Chalmers A, D'Cruz D, Wallace D et al (2008) The efficacy and safety of abatacept in SLE: Result of a 12-month exploratory study. American College of Rheumatology Annual Meeting, Abstract L 15

42   Daoussis D, Andonopoulos AP, Liossis SN (2004) Targeting CD40L: A promising therapeutic approach. *Clin Diagn Lab Immunol* 11: 635–41

43   Toubi E, Shoenfeld Y (2004) The role of CD40-CD154 interactions in autoimmunity and the benefit of disrupting this pathway. *Autoimmunity* 37: 457–64

44   Boumpas DT, Furie R, Manzi S, Illei GG, Wallace DJ, Balow JE et al (2003) A short course of BG9588 (anti-CD40 ligand antibody) improves serologic activity and decreases hematuria in patients with proliferative lupus glomerulonephritis. *Arthritis Rheum* 48: 719–27

45   Kalunian KC, Davis JC Jr, Merrill JT, Totoritis MC, Wofsy D (2002) Treatment of systemic lupus erythematosus by inhibition of T cell costimulation with anti-CD154: A randomized, double-blind, placebo-controlled trial. *Arthritis Rheum* 46: 3251–8

46   Rodriguez-Palmero M, Franch A, Castell M, Pelegri C, Perez-Cano FJ, Kleinschnitz C et al. (2006) Effective treatment of adjuvant arthritis with a stimulatory CD28-specific monoclonal antibody. *J Rheumatol* 33: 110–8

47   Suntharalingam G, Perry MR, Ward S, Brett SJ, Castello-Cortes A, Brunner MD et al (2006) Cytokine storm in a phase 1 trial of the anti-CD28 monoclonal antibody TGN1412. *N Engl J Med* 355: 1018–28

Immunobiology of IL-6 – Tocilizumab (humanised anti-IL-6 receptor antibody) for the treatment of rheumatoid arthritis

Yoshiyuki Ohsugi[1] and Tadamitsu Kishimoto[2]

[1]Chugai Pharmaceutical Co. Ltd., Tokyo 103-8324, Japan
[2]Graduate School of Frontier Bioscience, Osaka University, Osaka 565-0871, Japan

Abstract

The cloning of IL-6 cDNA in 1986 revealed that IL-6 is a multifunctional cytokine that plays important roles in the immunopathogenesis of rheumatoid arthritis (RA). A close relationship was observed between IL-6 levels in the synovial compartment and disease activity in RA patients, and overproduction of IL-6 could readily explain the abnormal laboratory findings and clinical symptoms seen in these patients. IL-6 therefore appeared to be a worthwhile and attractive therapeutic target for RA. In practice, blockage of IL-6 signalling by a humanised anti-IL-6 receptor antibody [tocilizumab (TCZ); also known as MRA] has been found to be very effective in the treatment of patients with RA. In recent Japanese Phase III clinical studies in RA patients, TCZ clearly prevented radiographic progression of joint destruction and greatly improved signs and symptoms. Very interestingly and importantly, this therapy has also proved quite effective at improving fever, fatigue and anaemia. No serious adverse events have been reported. At present, several international clinical studies of TCZ are ongoing in more than 4000 patients with active RA in 41 countries. The results are continuing to confirm the efficacy and safety of TCZ in the treatment of patients with RA.

Introduction

IL-6 was originally identified as a T cell-derived soluble factor that causes differentiation of B cells into antibody-producing plasma cells [1]. When the gene coding for IL-6 was cloned in 1986, it became apparent that IL-6 had been studied under several different names in various laboratories. It has since been clarified that IL-6 does indeed have various biological functions in addition to B cell activation, and it is now well known that IL-6 plays important roles in immunity, inflammation and haematopoiesis. More importantly, evidence has accumulated that deregulation of IL-6 results in the development of various autoimmune diseases. In fact, the multiple biological activities of IL-6 provide explanations for various clinical symptoms of rheumatoid arthritis (RA) [2–4].

These findings suggest that IL-6 is a worthwhile and attractive therapeutic target molecule for RA. In this review, we discuss the immunopathological roles of IL-6 in RA, and the clinical usefulness of tocilizumab (TCZ), a humanised anti-IL-6R antibody that blocks IL-6 signalling, in the treatment of RA.

IL-6 has various biological activities

IL-6 acts on various cell types and has a variety of biological functions, e.g. IL-6 acts as a hepatocyte-stimulating factor (HSF) [5, 6]. Acute inflammation is accompanied by changes in the plasma concentration of many proteins, such as a decrease in albumin and increases in many "acute-phase proteins", including C-reactive protein (CRP), fibrinogen, serum amyloid A protein and haptoglobin. Inflammation, injury and cancer all induce the expression of IL-6, resulting in increased synthesis of acute-phase proteins in the liver [5, 6]. Moreover, in IL-6 knockout mice, it has been shown that IL-6 is essential for antiviral antibody response, as well as for the induction of acute-phase reaction [7]. We have also reported that the injection of recombinant human IL-6 into cynomolgus monkeys increased the serum CRP level and the platelet count in the peripheral blood [8].

One of the recent advances regarding IL-6 is the discovery that IL-6 plays a critical role in the development of chronic anaemia. It has been shown that IL-6 induces hepcidin, which is an iron regulatory peptide produced in the liver. Hepcidin regulates the recycling of iron by macrophages and the absorption of iron from the intestine. Thus, excessive IL-6 causes hypoferraemia, which leads to "anaemia of chronic inflammation" (also known as "anaemia of chronic disease").

Another important activity of IL-6 is the induction of osteoclast differentiation, which may contribute to joint destruction in patients with RA. IL-6 also stimulates the expression of vascular endothelial growth factor (VEGF), which is an essential factor for neo-vascularisation. Moreover, IL-6 also enhances the function of leptin, an anti-appetite hormone, resulting in anorexia in patients with chronic inflammatory diseases. In addition, it has been reported that injection of IL-6 into cancer patients caused fever.

IL-6 signalling pathway: The IL-6 receptor and gp130

We succeeded in isolating the cDNA for IL-6 and IL-6 receptor in 1986 and 1988, respectively [9, 10]. We found that the receptor has an Ig-like domain at the N terminus but no unique sequences in any other regions. It also has a very short intra-cytoplasmic portion and no kinase domains. These features make it unlike what is considered an "authentic receptor". Another protein, a 130 kDa cell-surface glycoprotein that we named gp130, is necessary for IL-6 signal transduction. We isolated a cDNA encoding gp130. It was eventually concluded that the full IL-6 receptor consists of two polypeptide chains of 80 and 130 kDa, and that IL-6 stimulation triggers association of these two chains leading to IL-6 signalling [11–14]. A recent crystal structure study has demonstrated that two of each molecule associate to form a hexamer complex [15, 16].

Importantly, gp130 is expressed ubiquitously in all tissues, even in cells that lack detectable expression of the 80-kDa IL-6 receptor [12]. This suggested that gp130 is not merely a component of IL-6 receptor, and that it might function as a common signal transducer for various cytokines. In fact, many different cytokines do share the same receptor component, and this can explain the redundant activity of several cytokines.

We and others have reported that ciliary neurotropic factor (CNTF), leukaemia inhibitory factor (LIF), oncostatin M (OM), IL-11 and cardiotropin-1 (CT-1) all use gp130 as a component of their receptors [17–20]. This explains why these cytokines have very similar activities.

More importantly, soluble IL-6 receptor (sIL-6R) that lacks transmembrane and cytoplasmic domains is present in serum and synovial fluids. Once sIL-6R binds to its ligand, the complex becomes capable of associating with gp130 to transduce the IL-6 signal into cells. This means that the IL-6 signalling pathway functions, by means of gp130, even for cells that do not express IL-6R on their surface. This is called trans-signalling (Fig. 1A).

IL-6 and RA

It became evident that IL-6 is involved in various diseases, including chronic inflammation. While trying to isolate the cDNA for IL-6, we noticed that the same activity was observed in cardiac myxoma cells [21, 22]. Cardiac myxoma is a benign heart tumour that arises from the atrium. Patients with cardiac myxoma exhibit a wide variety of autoimmune and inflammatory symptoms, including autoantibodies, fever, joint pains and anaemia. All these symptoms disappear after surgical removal of the tumour.

We found that cardiac myxoma cells produce a large amount of IL-6. This result suggested that IL-6 might contribute to the pathology of autoimmune diseases and play an important role, not only in B cell immunology, but also in a variety of disease symptoms.

We also found an abnormal overproduction of IL-6 in patients with Castleman's disease [23]. Affected lymph node cells overproduce IL-6, which explains symptoms such as high fever, anaemia, fatigue, anorexia, acute-phase reactions, hypergammaglobulinaemia, secondary amyloidosis and massive plasma cell infiltration into

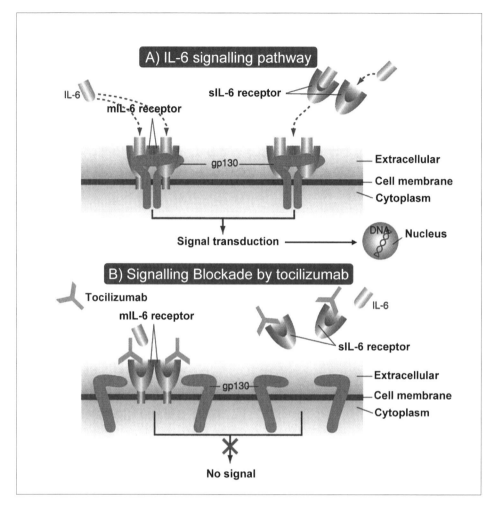

Figure 1
IL-6 signalling pathway and blockade of the signalling by tocilizumab (modified from: Bio-technology J. *Yodosha (2006) 7–8: 517–520).*

affected lymph nodes. In 1988, we reported constitutive overproduction of IL-6 by synovial tissues in RA patients [24]. This readily explains almost all the symptoms seen in RA patients. Consistent with this notion, there is a highly significant correlation between IL-6 levels in the RA synovium and scores for local disease activity [25]. Based on the above findings in patients with cardiac myxoma, RA and Castleman's disease, we concluded that the blockade of IL-6 and its receptor interactions was a promising new therapeutic approach for the treatment of these diseases.

From laboratory to clinic: IL-6R as a therapeutic target for RA

Research and development of a recombinant humanised anti-IL-6R antibody

On the basis of the above experimental and clinical results, we set out to develop an anti-IL-6 receptor blockade therapy (Fig. 1B). In collaboration with the MRC Collaborative Centre in London, mouse monoclonal antibody-binding human IL-6 receptor was humanised by means of complementarity-determining region (CDR) grafting technology [26].

Clinical response in patients with Castleman's disease

The humanised antibody was administered to seven patients with multicentric Castleman's disease, with the approval of the institute's ethical committee and the patients' consent [27]. Immediately after the antibody was administered, fever and fatigue disappeared, while anaemia and serum levels of CRP, fibrinogen and albumin started to improve. After 3 months of treatment, hypergammaglobulinaemia and lymphadenopathy were remarkably alleviated, as were the renal function abnormalities in the patients with amyloidosis. The pathophysiological significance of IL-6 in Castleman's disease was thus confirmed, and blockade of IL-6 signalling by the anti-IL-6 receptor antibody was shown to be a potential new therapy for IL-6-related diseases. A Phase II clinical trial in 28 patients with Castleman's disease was conducted in 2002. The antibody showed significant effects in all patients [28], so it was launched as a therapeutic drug for Castleman's disease (an orphan disease) in 2005.

IL-6 is a critical cytokine in experimental arthritis in animal models

To investigate the direct role of IL-6 in the development of RA, IL-6-deficient mice were backcrossed into C57BL/6 mice for eight generations, and the histological manifestations following the induction of antigen-induced arthritis in IL-6-deficient mice and wild-type littermates were compared [29]. Wild-type mice developed severe arthritis, whereas IL-6-deficient mice displayed little or no arthritis. The expression of TNF mRNA in synovial tissues in IL-6-deficient mice was comparable to that in wild-type mice, even though no arthritis was observed in the former. Recently, S. Sakaguchi and colleagues reported that deleting the IL-6 gene in the SKG mice (which develop RA owing to a mutation on the T cell signalling pathway) gave complete protection from development of RA, whereas 20% of TNF-α-deficient SKG mice developed the disease [30, 31]. All of these basic studies encouraged us to apply anti-IL-6 receptor therapy to patients with RA.

Efficacy and safety of TCZ in the treatment of RA

Following the success of the experimental treatment of Castleman's patients, TCZ was applied to the treatment of RA. The first step was to conduct Phase I/II trials of tocilizumab in Japan [32] and the UK [33]. The pharmacokinetics of TCZ were also investigated, especially in the Japanese Phase I/II trial.

The Japanese trial was an open-label, multi-dose study in 15 RA patients. TCZ (2, 4 or 8 mg/kg body weight) was administered to patients with active RA every 2 weeks for 6 weeks. Serum CRP and serum amyloid A (SAA) levels were completely normalised as long as TCZ was detectable in the serum, indicating that IL-6 is essential for the production of CRP and SAA *in vivo*. As a result, the levels of these acute-phase proteins could be used as surrogate markers to indicate whether the TCZ concentration was adequate to block IL-6 activity.

The UK trial was a single-dose, randomised, double-blind, placebo-controlled dose-escalation trial in 45 patients with active RA. These patients received a single intravenous dose of 0.1, 1, 5 or 10 mg/kg TCZ or placebo.

In both trials, TCZ was well tolerated and no serious adverse events were observed. That encouraged us to move into Phase II trials of TCZ in RA patients.

Phase II trials of TCZ in patients with RA

The safety and efficacy of TCZ treatment were evaluated in multi-centre, double-blind, randomised, placebo-controlled Phase II trials in RA patients in Japan [34] and Europe [35] that were completed in 2001 and 2002, respectively. In Japan, 164 patients with refractory RA received 4 or 8 mg/kg TCZ, or placebo, i.v. every 4 weeks for a total of 12 weeks, and the clinical response was evaluated using the American College of Rheumatology (ACR) criteria. As reported by Nishimoto et al. [34], TCZ treatment significantly improved all measures of disease activity in the ACR core set, and the results were comparable to or better than those obtained with anti-TNF antibody or soluble TNF receptor therapy. The incidence of at least 20% improvement in disease activity according to the ACR criteria (ACR20) was 78% in the higher dose group. This was higher than in the lower-dose group (57%), and significantly higher than in the control group (11%) ($p < 0.001$). The incidences of 50% and 70% improvements in disease activity (ACR50 and ACR70) were 40% and 16%, respectively, in the higher dose group, and both of these incidences were significantly higher than in the placebo group. Efficacy was also evaluated using the Disease Activity Score 28-joint count (DAS28) categories; the incidence of good or moderate response was 91% in the 8 mg/kg group, 72% ($p = 0.012$) in the 4 mg/kg group and 19% ($p < 0.001$) in the placebo group. TCZ treatment also improved laboratory findings such as haemoglobin levels, platelet counts, and the serum levels of CRP, fibrinogen, SAA, albumin and rheumatoid factors. In addition, TCZ treatment significantly improved bone metabolism, suggesting that IL-6 blockade may prevent the osteoporosis seen in RA patients. In long-term trials (more than 15 months), ACR20, 50 and 70 reached 88%, 67% and 42%, respectively. During long-term administration, the serum IL-6 level gradually decreased, becoming undetectable in some patients. This suggests that anti-IL-6 receptor therapy may go beyond simple anti-inflammatory therapy to affect fundamental aspects of the immune system.

In relation to safety, the overall incidences of adverse events were 56%, 59% and 51% in the placebo, 4 and 8 mg/kg groups, respectively, so there was no dose dependency in adverse events. One patient died because of reactivation of chronic active Epstein-Barr virus (EBV) infection and consequent haemophagocytosis syndrome after receiving a single dose of 8 mg/kg TCZ. Retrospectively, it was found that she had Hodgkin's disease with increased EBV DNA in plasma before enrolment in the study, but had not been excluded [36].

Among laboratory findings, an increase in total cholesterol level was reported frequently (44% of patients) in the TCZ groups. However, mean total cholesterol levels did not continue to increase with repeated dosing, and stabilised close to the upper limit of the normal range. High-density lipoprotein (HDL) cholesterol levels also increased, so the atherogenic index [(total cholesterol – HDL cholesterol)/HDL cholesterol] did not change throughout the study period. No cardiovascular complications associated with increased total cholesterol were observed.

Mild to moderate increases in liver function test values were also observed in 14 (12.8%) of 109 patients in the TCZ groups. The above data indicate that TCZ treatment is generally well tolerated and shows clinical benefits.

In a Phase II study conducted in Europe (the CHARISMA study) [35], 359 patients with active RA and an inadequate response to methotrexate (MTX) therapy (≥10 mg/week MTX for at least 6 months) were administered TCZ or TCZ placebo together with 10–25 mg/week of MTX or MTX placebo every 4 weeks for a total of 12 weeks.

The patients were randomised to receive 2, 4 or 8 mg/kg TCZ, either as monotherapy or in combination with MTX, or MTX monotherapy. As evaluated by change in DAS28 from baseline, 8 mg/kg TCZ monotherapy and 8 mg/kg TCZ plus MTX both yielded significantly higher responses than MTX alone. However, there was no significant difference between 8 mg/kg TCZ monotherapy and 8 mg/kg TCZ plus methotrexate.

Phase III trials of TCZ in patients with RA

Large-scale Phase III trials have been completed in Japan and other countries, including Europe and the United States, and the results have recently been published.

In Japan, a Phase III, randomised, controlled trial was performed to investigate the efficacy and safely of TCZ treatment in 306 patients with active RA [37].

Patients with disease duration of less than 5 years were randomised to receive either TCZ monotherapy (8 mg/kg i.v. once every 4 weeks) or conventional disease modifying anti-rheumatic drugs (DMARDs) for 52 weeks.

As measured by change in Total Sharp Score, patients in the TCZ group showed a significant delay in the radiographic progression of joint destruction compared with those receiving conventional DMARDs (mean values: 2.3 vs 6.1) ($p < 0.01$).

Compared with the placebo group, TCZ also significantly decreased erosion and joint space narrowing ($p < 0.001$ and $p < 0.018$, respectively). In relation to signs and symptoms, ACR20, 50, and 70 response was achieved by 89%, 70% and 47%, respectively, of patients in the TCZ group, which was significantly better than the 35%, 14% and 6% who achieved these responses in the conventional DMARD group ($p < 0.001$) (Fig. 2).

The overall incidences of adverse events were 89% and 82% in the TCZ and control groups, respectively. This trial showed that TCZ monotherapy is more efficacious than conventional DMARDs at delaying and stopping radiographic progression of joint destruction, and at improving signs and symptoms [37].

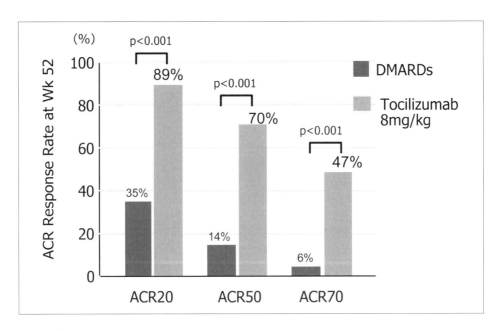

Figure 2
Results of the Japanese Phase III trial in RA patients: Improvements in signs and symptoms
[37]. Patients received tocilizumab at a dose of 8 mg/kg or placebo every 4 weeks. ACR
response rate was compared at 52 weeks. Statistical difference was analysed by paired
t-test.

The results of the first two of five multi-national Phase III studies have provided further evidence that IL-6 receptor inhibition is likely to play a significant role in the treatment of RA. In the first multi-national Phase III study (the OPTION study, a double-blind, randomised, controlled study), 623 patients with moderate to severe active RA, refractory to MTX, were allocated to receive 8 or 4 mg/kg TCZ, or placebo, i.v. every 4 weeks. All three groups also received MTX. The results show that the proportion of patients that achieved ACR20 response at 24 weeks was significantly higher in the 8 and 4 mg/kg TCZ groups than in the placebo group ($p < 0.0001$). Reduction in disease activity score (DAS-28) was observed in both TCZ groups. The proportion of patients who achieved a good or moderate EULAR response at 24 weeks was also significantly higher in the TCZ groups than in the placebo group ($p < 0.0001$). These data demonstrate that TCZ is highly effective and has a good safety and tolerability profile [38].

The second multi-national Phase III study [the TOWARD (tocilizumab in combination with traditional DMARD therapy) study] was conducted in 1216 patients with moderate to severe active RA and inadequate response to DMARDs. The study was conducted at 130 study sites in 18 countries, including the United States. In this two-arm, randomised, double-blind study, patients received either 8 mg/kg TCZ or placebo i.v. every 4 weeks in combination with stable anti-rheumatic therapy, including traditional DMARDs but excluding biologicals. Compared to patients treated with traditional DMARDs alone, a greater proportion of patients treated with TCZ plus traditional DMARDs achieved significant improvement in disease signs and symptoms at 24 weeks. This study also explored the pharmacokinetics and pharmacodynamic parameters of TCZ, as well as immune response to TCZ, in this patient population. Patient symptoms were measured using the standard ACR score assessment method. The TOWARD trial data further document the efficacy and safety of TCZ and the value of its IL-6 receptor inhibiting activity (Press release by Roche). The results were presented at ACR meetings in 2007.

The above clinical studies have clearly indicated the benefits of using TCZ to block IL-6 signalling in the treatment of patients with RA.

Possible mechanisms of action of TCZ

The fact that the effects of TCZ therapy in RA patients have been so dramatic suggests that the overproduction of IL-6 is deeply involved in the pathogenesis and progression of RA. As mentioned above, excessive production of IL-6 can readily explain almost all of the symptoms seen in RA patients.

For example, blockade of IL-6 signalling by TCZ causes dramatic improvement of anaemia, which is highly beneficial for maintaining and improving the quality of life of RA patients. This effect could be the result of the inhibition of the production of hepcidin (an iron regulatory peptide secreted by the liver cells) [39–42] or the recovery of signal transduction *via* the erythropoietin (EPO) receptor. The EPO receptor and IL-6R share the Janus kinase-signal transducer and activator of transcription (JAK-STAT) signalling pathway [43]. Since excessive IL-6 signalling induces expression of suppressors of cytokine signalling (SOCS), which are intracellular negative feedback factors that inhibit the JAK-STAT pathway, TCZ may down-regulate these factors, resulting in increased EPO signalling over time.

Another mechanism of the activity of TCZ may be that blockade of IL-6 signalling causes a decrease in serum VEGF, which inhibits angiogenesis in the synovial tissues, which inhibits hyperplasia of the synovium [44].

Yet another possible mechanism is that interference with activation/differentiation of osteoclasts may contribute to the prevention of joint destruction [45]. Finally, one of the most notable recent advances is related to the discovery that a newly identified type of T helper cell, the Th17 cell (which produces IL-17), may be deeply involved in the pathogenesis of autoimmune diseases, including RA [46–48]. Very excitingly, IL-6, together with TGF-β, is involved in the differentiation of this particularly pathogenic T cell lineage, and it has already been found that blockade of IL-6 signalling results in suppression of the development of Th17 cells in mice [49]. Thus, it seems that TCZ is far more than just an anti-inflammatory agent, and it inhibits the pathogenesis of RA by its effects on the underlying aetiology of the disease.

Further comparisons of the immune system and gene expression before and after TCZ treatment in RA patients may provide important insights into the pathogenesis of RA. If so, the transition "from laboratory to clinic" will lead to a further transition "from clinic to basic studies".

References

1   Kishimoto T (1989) The biology of interleukin-6. *Blood* 74: 1–10
2   Kishimoto T (2005) Interleukin-6: From basic science to medicine-40 years in immunology. *Annu Rev Immunol* 23: 1–21
3   Gershwin ME, Ohsugi Y (eds) (2005) The immunobiology of IL-6. *Clin Rev Allergy Immunol* 28: 177–269
4   Nishimoto N, Kishimoto T (2006) Interleukin 6: from bench to bedside. *Nat Clin Pract Rheumatol* 2: 619–626
5   Gauldie J, Richards C, Harnish D et al (1987) Interferon β/B-cell stimulatory factor type 2 shares identity with monocyte-derived hepatocyte-stimulating factor and regulates the major acute phase protein response in liver cells. *Proc Natl Acad Sci USA* 84: 7251–55
6   Andus T, Geiger T, Hirano T et al (1987) Recombinant human B cell stimulatory factor 2 (BSF-2/IFN-β2) regulates β-fibrinogen and albumin mRNA levels in Fao-9 cells. *FEBS Lett* 221: 18–22

7   Kopf M, Baumann H, Freer G et al (1994) Impaired immune and acute-phase responses in interleukin-6-deficient mice. *Nature* 368: 339–4

8   Imazeki I, Saito H, Hasegawa M et al (1998) IL-6 functions in cynomolgus monkeys blocked by a humanized antibody to human IL-6 receptor. *Int J Immunopharmacol* 20: 345–357

9   Hirano T, Yasukawa K, Harada H et al (1986) Complementary DNA for a novel human interleukin (BSF-2) that induces B lymphocytes to produce immunoglobulin. *Nature* 324: 73–76

10 Yamasaki K, Taga T, Hirata Y et al (1988) Cloning and expression of the human interleukin-6 (BSF-2/IFN beta 2) receptor. *Science* 241: 825–828

11 Taga T, Hibi M, Hirata Y et al (1989) Interleukin-6 triggers the association of its receptor with a possible signal transducer, gp130. *Cell* 58: 573–581

12 Hibi M, Murakami M, Saito M et al (1990) Molecular cloning and expression of an IL-6 signal transducer, gp130. *Cell* 63: 1149–57

13 Murakami M, Narazaki M, Hibi M et al (1991) Critical cytoplasmic region of the interleukin 6 signal transducer gp130 is conserved in the cytokine receptor family. *Proc Natl Acad Sci USA* 88: 11349–53

14 Murakami M, Hibi M, Nakagawa N et al (1993) IL-6-induced homodimerization of gp130 and associated activation of a tyrosine kinase. *Science* 260: 1808–10

15 Varghese JN, Moritz RL, Lou MZ et al (2002) Structure of the extracellular domains of the human interleukin-6 receptor α-chain. *Proc Natl Acad Sci USA* 99: 15959–64

16 Skiniotis G, Boulanger MJ, Garcia KC, Walz T (2005) Signaling conformations of the tall cytokine receptor gp130 when in complex with IL-6 and IL-6 receptor. *Nat Struct Mol Biol* 12: 545–51

17 Ip NY, Nye SH, Boulton TG et al (1992) CNTF and LIF act on neuronal cells *via* shared signalling pathways that involve the IL-6 signal transducing receptor component gp130. *Cell* 69: 1121–32

18 Gearing DP, Comeau MR, Friend DJ et al (1992) The IL-6 signal transducer, gp130: An oncostatin M receptor and affinity converter for the LIF receptor. *Science* 255: 1434–37

19 Liu J, Modrell B, Aruffo A et al (1992) Interleukin-6 signal transducer gp130 mediates oncostatin M signaling. *J Biol Chem* 267: 16763–66

20 Yin T, Taga T, Tsang ML et al (1993) Involvement of IL-6 signal transducer gp130 in IL-11–mediated signal transduction. *J Immunol* 151: 2555–61

21 Hirano T, Taga T, Nakano N et al (1985) Purification to homogeneity and characterization of human B-cell differentiation factor (BCDF or BSFp-2). *Proc Natl Acad Sci USA* 82: 5490–94

22 Jourdan M, Bataille R, Seguin J et al (1990) Constitutive production of interleukin-6 and immunologic features in cardiac myxomas. *Arthritis Rheum* 33: 398–402

23 Yoshizaki K, Matsuda T, Nishimoto N et al (1989) Pathogenic significance of interleukin-6 (IL-6/BSF-2) in Castleman's disease. *Blood* 74: 1360–67

24   Hirano T, Matsuda T, Turner M et al (1988) Excessive production of interleukin 6/B cell stimulatory factor-2 in rheumatoid arthritis. *Eur J Immunol* 18: 1797–801

25   Tak PP, Smeets TJ, Daha MR et al (1997) Analysis of the synovial cell infiltrate in early rheumatoid synovial tissue in relation to local disease activity. *Arthritis Rheum* 40: 217–25

26   Sato K, Tsuchiya M, Saldanha J et al (1993) Reshaping a human antibody to inhibit the interleukin-6-dependent tumor cell growth. *Cancer Res* 53: 851–6

27   Nishimoto N, Sasai M, Shima Y et al (2000) Improvement in Castleman's disease by humanized anti-interleukin-6 receptor antibody therapy. *Blood* 95: 56–61

28   Nishimoto N, Kanakura Y, Aozasa K et al (2005) Humanized anti-interleukin-6 receptor antibody treatment of multicentric Castleman's disease. *Blood* 106: 2627–32

29   Ohshima S, Saeki Y, Mima T et al (1998) Interleukin 6 plays a key role in the development of antigen-induced arthritis. *Proc Natl Acad Sci USA* 95: 8222–26

30   Sakaguchi N, Takahashi T, Hata H et al (2003) Altered thymic T-cell selection due to a mutation of the ZAP-70 gene causes autoimmune arthritis in mice. *Nature* 426: 454–60

31   Hata T, Sakaguchi N, Yoshitomi H et al (2004) Distinct contribution of IL-6, TNF-α, IL-1, and IL-10 to T cell-mediated spontaneous autoimmune arthritis in mice. *J Clin Invest* 114: 582–88

32   Nishimoto N, Yoshizaki K, Maeda K et al (2003) Toxicity, pharmacokinetics, and dose finding study of repetitive treatment with humanized anti-interleukin 6 receptor antibody, MRA, in rheumatoid arthritis – Phase I/II clinical study. *J Rheumatol* 30: 1426–35

33   Choy EH, Isenberg DA, Garrood T et al (2002) Therapeutic benefit after blocking interleukin-6 activity in rheumatoid arthritis with an anti-interleukin-6 receptor monoclonal antibody. *Arthritis Rheum* 46: 3143–50

34   Nishimoto N, Yoshizaki K, Miyasaka N et al (2004) Treatment of rheumatoid arthritis with humanized anti-interleukin-6 receptor antibody: a multicenter, double-blind, placebo-controlled trial. *Arthritis Rheum* 50: 1761–69

35   Maini RN, Taylor PC, Szechinski J et al (2006) Double-blind randomized controlled clinical trial of the interleukin-6 receptor antagonist, tocilizumab, in European patients with rheumatoid arthritis who had an incomplete response to methotrexate. *Arthritis Rheum* 54: 2817–29

36   Ogawa J, Harigai M, Akashi T et al (2006) Exacerbation of chronic active Epstein-Barr virus infection in a patient with rheumatoid arthritis receiving humanised anti-interleukin-6 receptor monoclonal antibody. *Ann Rheum Dis* 65: 1667–9

37   Nishimoto N, Hashimoto J, Miyasaka N et al (2007) Study of active controlled monotherapy used for rheumatoid arthritis, an IL-6 inhibitor (SAMURAI): Evidence of clinical and radiographic benefit from an X-ray reader-blinded randomized, controlled trial of tocilizumab. *Ann Rheum Dis* 66: 1162–7

38   Smolen JS, Beaulieu A, Rubbert-Roth A et al (2008) Effect of interleukin-6 inhibition

with tocilizumab in patients with rheumatoid arthritis (OPTION study) A double-blind, placebo-controlled, randamized trial. *Lancet* 371: 987–97

39 Nemeth E, Valore EV, Territo M et al (2003) Hepcidin, a putative mediator of anemia of inflammation, is a type II acute-phase protein. *Blood* 101: 2461–3

40 Nemeth E, Rivera S, Gabayan V et al (2004) IL-6 mediates hypoferremia of inflammation by inducing the synthesis of the iron regulatory hormone hepcidin. *J Clin Invest* 113: 1271–6

41 Lee P, Peng H, Galbart T et al (2005) Regulation of hepcidin transcription by interleukin-1 and interleukin-6. *Proc Natl Acad Sci USA* 102: 1906–10

42 Nemeth E, Ganz T (2006) Regulation of iron metabolism by hepcidin. *Annu Rev Nutr* 26: 323–42

43 Sasaki A, Yasukawa H, Shouda T et al (2000) CIS/SOCS-3 suppresses erythropoietin (EPO) signaling by binding the EPO receptor and JAK2. *J Biol Chem* 275: 29338–47

44 Nakahara H, Song J, Sugimoto M et al (2003) Anti-interleukin-6 receptor antibody therapy reduces vascular endothelial growth factor production in rheumatoid arthritis. *Arthritis Rheum* 48: 1521–9

45 Tamura T, Udagawa N, Takahashi N et al (1993) Soluble interleukin-6 receptor triggers osteoclast formation by interleukin-6. *Proc Natl Acad Sci USA* 90: 11924–8

46 Mangan PR, Harrington LE, O'Quinn DB et al (2006) Transforming growth factor-_ induces development of the T_H17 lineage. *Nature* 441: 231–234

47 Bettelli E, Carrier Y, Gao W et al (2006) Reciprocal developmental pathways for the generation of pathogenic effector T_H17 and regulatory T cells. *Nature* 441: 235–238

48 Veldhoen M, Hocking RJ, Atkins CJ et al (2006) TGFbeta in the context of an inflammatory cytokine milieu supports *de novo* differentiation of IL-17-producing T cells. *Immunity* 24: 179–189

49 Kimura A, Naka T, Kishimoto T et al (2007) IL-6-dependent and independent pathways in the development of interleukin 17-producing T helper cells. *Proc Natl Acad Sci USA* 104: 12099–104

Role of IL-1 in erosive arthritis, lessons from animal models

Wim B. van den Berg, Leo A. B. Joosten and Fons A. J. van de Loo

Rheumatology Research & Advanced Therapeutics, Department of Rheumatology, Radboud University Nijmegen Medical Centre, Geert Grooteplein 28, 6525 GA, Nijmegen, The Netherlands

Abstract

Tumor necrosis factor (TNF), interleukin-1 (IL-1) and IL-6 are considered master cytokines in chronic destructive arthritis. IL-1 drives chronic erosive arthritis and its blockade has been shown to ameliorate joint destruction in many animal models of arthritis. This ranges from a dominant role of IL-1 in immune complex arthritis, to a key role in development of T cell-dependent arthritis and TNF transgenic arthritis. This makes IL-1 an attractive therapeutic target, in addition to TNF and IL-6. However, IL-1 dependency can be lost under conditions of T cell IL-17 abundance as well as the presence of Toll-like receptor ligands. The latter may underlie the variable responsiveness of rheumatoid arthritis patients to anti-cytokine therapy and warrants combination therapy for optimal control.

Introduction

Studies in well-defined animal models of arthritis make it clear that tumor necrosis factor (TNF) is involved in early joint swelling and cell influx. However, TNF alone is poorly arthritogenic and hardly destructive, and exerts its full arthritogenic potential through induction of IL-1. Intriguingly, TNF-independent IL-1 production is found in many model situations, including pathways driven by macrophages, T cells and immune complexes. Its relevance is underlined by the great efficacy of anti-IL-1 therapy and a profound lack of erosive arthritis in IL-1β-deficient mice. IL-1 is a prominent inducer of RANKL and RANKL-mediated activation of osteoclasts. TNF, in synergy with T cell-derived IL-17 also up-regulates RANKL and induces bone erosion. Cartilage destruction is heavily dependent on IL-1. IL-1 is a strong activator of chondrocytes, induces cartilage breakdown through up-regulation of metalloproteinases and causes profound suppression of cartilage matrix synthesis. This catabolic activity, combined with impaired anabolic activity, results in marked cartilage loss. Collagen damage and therefore irreversible cartilage erosion is greatly amplified by the presence of immune complexes in the joint, through Fcγ receptor-mediated activation of IL-1-induced latent metalloproteinases [1]. Cartilage destruc-

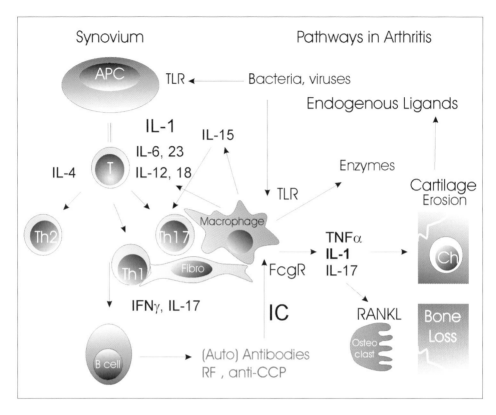

Figure 1
The vicious circle in chronic joint inflammation. Schematic presentation of pathways of syno-
vitis and concomitant cartilage and bone destruction. Note the amplifying elements through
endogenous TLR ligands, T cell activation and generation of autoantibodies. The latter will
trigger macrophages after immune-complex (IC) formation, through Fcγ receptors. IL-1 plays
an important role in enhancing T cell activation through induction of CD40L and OX40.
Thus, excess IL-1 signaling may activate these pathways, leading to the development of T
cell-mediated autoimmune diseases. TLR, Toll-like receptor; Ch, chondrocyte.

tion induced by IL-1 can itself further amplify and perpetuate joint inflammation
in two ways. Breakdown fragments such as biglycan provide endogenous activators
for Toll-like receptors (TLRs), in particular TLR4, on synovial macrophages and
fibroblasts, leading to inflammatory mediator production [2]. In addition, these
fragments may form autoimmune stimuli and may sustain arthritis when tolerance
is lost. IL-1, together with IL-6, is probably a major driver of Th17 generation
against autologous antigens. Some elements of the pathways discussed are depicted
in Figure 1.

IL-1 synthesis, activation and signaling pathway

The IL-1 family consists of ten members, including IL-1α, IL-1β, IL-1 receptor antagonist (IL-1Ra) and IL-18 [3–5]. IL-1α and IL-1β are produced from two different genes located on chromosome 2 and synthesized as 31-kDa precursors. Pro-IL-1α is a cell-bound cytokine and is activated by proteases called calpains. There is growing evidence that IL-1α is involved in intracellular signaling [6]. The production of IL-1β is *via* non-classical pathways of protein secretion. TLR agonists such as endotoxins initiate the synthesis of the inactive IL-1β precursor. The IL-1β precursor co-localizes with procaspase-1 followed by the conversion of the inactive procaspase-1 to active caspase-1 by a complex of proteins termed the "IL-1β inflammasome" (Fig. 2) [6, 7]. In resting cells procaspase-1 is bound to a large inhibitor molecule, which prevents its activation. During initiation of IL-1β synthesis, there is activation of caspase-1, which then processes the IL-1β precursor into a mature form ready for secretion. Autocatalytic activation of pro-caspase-1 occurs *via* efflux of potassium ions as a result of triggering the P2X7 receptor by ATP [6]. Recently, it was found that a small peptide LL37 released from activated neutrophils and epithelial cells can stimulate the secretion of mature IL-1β *via* the P2X7 receptor [8]. There is evidence that under inflammatory conditions IL-1β processing can be caspase-1 independent, hampering caspase inhibition as a therapeutic control. Using caspase-1 gene-deficient mice it was demonstrated that IL-1β was still produced in both acute and chronic joint inflammation. Several proteases, such as proteinase 3 and granzyme A, have been suggested to be involved in the caspase-1-independent cleavage of pro-IL-1β [9].

IL-1 binds to the IL-1R complex that consists of IL-1 receptor type I (IL-1R) and IL-1 receptor accessory protein (IL-1RacP). IL-1Ra is the natural inhibitor that is able to block IL-1R interaction. After binding of IL-1 to the IL-1R, IL-1RacP is recruited and a functional high-affinity complex is formed. Signaling occurs *via* MyD88 adaptor protein that binds to the TIR domain in the intracellular part of the receptor complex. The presence of the death domain in MyD88 allows recruitment of the IL-1 receptor-activated kinases (IRAK1-4). Thereafter, another adaptor protein is bound (TRAF-6), which leads to activation of several protein kinases [10], including JNK, ERKs and IKK (Fig. 3). Finally, this results in activation of transcription factors (NF-κB and AP-1) involved in regulation of inflammation-related gene expression, such as cytokines and chemokines.

IL-1 inhibition

Inhibition of IL-1 signaling can be achieved at various levels. Pivotal downstream elements are obvious therapeutic targets, but the specificity may be limited. Apart from the signaling IL-1R type I, a nonsignaling decoy receptor (IL-1R type II) was

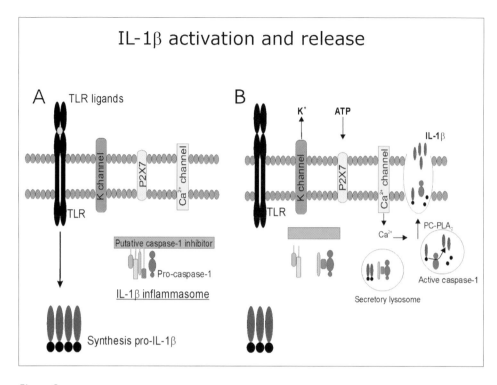

Figure 2

IL-1β synthesis, processing and secretion. (A) Gene expression and protein synthesis of IL-1β precursor is induced by TLR ligands such as endotoxin. The precursor remains in the cytosol of the cells together with inactive pro-caspase-1 that is bound to the IL-1β inflammasome complex. The inflammasome complex contains products of the NALP-3 gene and is in an inactive state due to binding to a putative inhibitor. (B) TLR signaling activates the inflammasome by uncoupling of the inhibitor and the NAPL-3 gene products from the pro-caspase-1.

identified, which has a natural regulatory function by binding IL-1 and consuming IL-1RacP. Soluble forms have been found of both the type II receptor and the IL-1RacP. An engineered form of the latter has been shown to be effective in collagen-induced arthritis (CIA), when applied with gene therapy [11].

The most studied and therapeutically applied inhibitor in rheumatoid arthritis (RA) patients is the natural receptor antagonist IL-1Ra. To fully prevent IL-1 signaling, excessive levels (> 1000-fold) of IL-1Ra are needed, which impairs therapeutic potential. Stabilized forms with better pharmacokinetics have been prepared and have shown efficacy in reduction of erosion in RA [12]. However, whether sufficient IL-1 blocking was reached is still a matter of debate. Neutralizing antibodies to IL-1

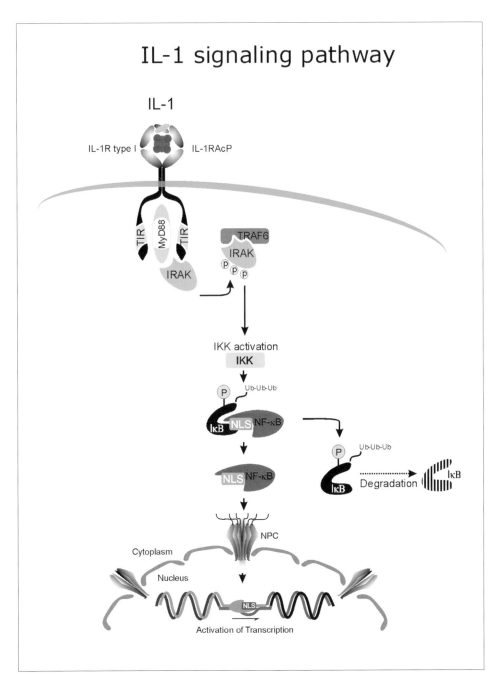

Figure 3
Intracellular IL-1 pathway.

are the most powerful tools for identifying the role of IL-1 in arthritis and have been used extensively in animal models. Therapeutic application is under study in recent trials in RA patients.

Arthritogenicity of IL-1

It is generally accepted that arthritis can be induced in mice by IL-1β. This was convincingly demonstrated by local injection of IL-1β or intra-articular overexpression by local gene transfer. One single injection of IL-1β in knee joints of mice results in disturbance of cartilage proteoglycan synthesis and influx of inflammatory cells [13]. Prolonged IL-1β exposure of rabbit or murine knee joints, using IL-1β gene transfer technology results in chronic destructive arthritis that resembles most features of RA. When compared with TNF-α, IL-1β is much more potent in inducing cartilage destruction *in vivo*. Tiny amounts of IL-1 are already sufficient to cause chondrocyte proteoglycan synthesis inhibition, whereas roughly a 100–1000-fold higher dose of TNF-α is needed to obtain the same effect [13]. IL-1β is the dominant form in most arthritis models (see below), but the potency of both isoforms is similar. Transgenic overexpression of human IL-1α results in florid arthritis, with a major role of membrane-bound IL-1 [14, 15].

A strong argument for the dominant role of IL-1 in erosive arthritis has emerged from studies in TNF transgenic mice. The group of George Kollias have already shown that arthritis was arrested when these mice were treated with antibodies against the IL-1R [16]. More recent work further clarified separate roles of TNF and IL-1 in inflammation and erosion. In TNF transgenic mice (hTNFtg) that were crossed with IL-1α,β-deficient mice, the synovial inflammation was almost unaffected. However, bone erosion was highly reduced and cartilage damage was absent [17]. TNF levels were still high, which implies that TNF alone is hardly erosive [18]. This has led to the conclusion that TNF-induced structural joint damage is mediated by IL-1.

Arguments for a role of IL-1 in erosive arthritis

Table 1 shows arguments for a leading role of IL-1 in destructive arthritis. Apart from a clear role in single mediator systems, the exact impact of IL-1 has been identified in models of arthritis using antibodies, IL-1Ra or IL-1 gene-deficient mice. It appears that considerable IL-1 production occurs in many models, independent of TNF. This is in line with greater efficacy of anti-IL-1 treatment as compared to anti-TNF treatment. TNF blockade was effective when started before or shortly after onset of CIA, whereas anti-IL-1 treatment was more efficient and also suppressed

Table 1 - Arguments for a dominant role of IL-1 in destructive arthritis.

High arthritogenic and erosive potency of IL-1

TNF-induced arthritis can be arrested with anti-IL-1R antibodies

Bone erosion is highly suppressed in hTNFtg/IL-1$^{-/-}$ mice

Cartilage erosion is fully prevented in hTNFtg/IL-1$^{-/-}$ mice

TNF-independent production of IL-1 in many models of arthritis

Greater efficacy of anti-IL-1 as compared with anti-TNF to reduce erosions in many models

Erosive arthritis models can still be induced in TNF$^{-/-}$ mice

No cartilage and bone destruction in models in IL-1$^{-/-}$ mice

IL-1Ra$^{-/-}$ mice, displaying uncontrolled IL-1 activity, develop arthritis

advanced erosive arthritis [19]. Studies in mice deficient for the TNF receptor or TNF itself showed reduced incidence and severity of CIA. However, once joints were affected, full progression to erosive damage was seen in an TNF-independent fashion [20, 21]. Similar examples are seen in other models and are described here under the respective headings.

First clinical trials showed major protective effects of anti-TNF treatment in RA patients. Although the initial experimental findings were in favor of IL-1 as compared to TNF, clinical studies in RA patients were unfortunately disappointing. The soluble IL-1 type I receptor used had a high affinity for IL-1Ra, thus scavenging the endogenous IL-1 inhibitor. Later studies with IL-1Ra as a therapeutic modality showed significant reduction of joint erosion, although the effects on joint inflammation were limited. Recently, trials were started with a solid neutralizing antibody to IL-1β, and have shown efficacy in a subgroup of RA [22].

Role of IL-1 in models of arthritis

Further insight into relative roles of TNF, IL-1 and IL-17 has emerged from detailed studies in a range of experimental arthritis models and findings are summarized in Table 2. The model systems include innate, nonimmune triggering of synovial cells, as well as different mixtures of arthritogenic pathways driven by T cells and immune complexes. Crucial findings are based on blocking studies with specific inhibitors as well as cytokine-deficient mice. Comparative studies with neutralizing antibodies are potentially flawed by the efficacy of the various antibodies to fully neutralize a particular cytokine. On the other hand, observations from knockout mice predominantly provide insight into a role at the onset of arthritis; however, studies with arthritis models in conditional cytokine knockouts are scant.

Table 2 - Involvement of IL-1 and other features of various murine arthritis models.

Collagen arthritis	greater effect of IL-1 blocking as compared to TNF
AIA	major effect of anti-IL-1 treatment on erosion
AIA-flare	shift to a dominant role of IL-17
Immune complex	more dominant role of IL-1 as compared to TNF
SCW arthritis	strong TNF dependence, role of IL-1 in cartilage damage
SCW rechallenge	shift from TNF to IL-1 to IL-17 after repeated flares
IL-1Ra$^{-/-}$ mice	excessive IL-1 allows generation of autoreactive Th17

Mixed T cell and immune complex pathways

CIA and antigen-induced arthritis (AIA) are models based on preimmunization with a cartilage-specific or an exogenous protein, with generation of T cell reactivity and antibodies. The onset of arthritis is a mixture of pathways driven by immune complexes and T cells. TNF is important at onset of CIA, but IL-1 blocking is highly efficacious both in acute and advanced stage [19]. The latter is probably linked to IL-1-mediated generation of cartilage-derived autoantigens, epitope spreading and a role of IL-1 in generation of T cell autoreactivity at the site. IL-1β is the dominant isotype, and IL-1Ra-deficient DBA mice show enhanced susceptibility [23]. IL-17 blockade was effective in established arthritis and mainly prevented erosions [24].

The onset of AIA is vigorous and only partly dependent on TNF and IL-1. Cartilage erosion and propagation of inflammation are dependent on IL-1 [25, 26]. Intriguingly, when smoldering chronic arthritis is exacerbated with a small dose of antigen, T cell-mediated flares were strongly IL-17 dependent [27], underlining that processes can become relatively TNF/IL-1 independent, when sufficient Th17 cells are generated at the site.

Immune complex arthritis

Arthritis induced by passive transfer of antibodies directed against a cationic antigen, planted in the joint showed some TNF dependency at onset, but arthritis could be completely blocked with anti-IL-1 antibodies [28]. More recent studies with autoimmune glucose-6-phosphate isomerase (GPI) antibodies were done in TNF and IL-1-deficient mice. Findings were similar. Arthritis incidence was reduced in TNF$^{-/-}$ mice, but some animals showed undisturbed, severe arthritis, indicating that TNF helps to set arthritis in motion, but is not crucial anymore in propagation of immune complex arthritis. In contrast, IL-1-deficient mice were strongly protected [29]. Mast cells contribute to immune complex arthritis expression through release of IL-1 [30].

T cell-driven arthritis

The classic model of adjuvant arthritis (AA) in rats is a pure T cell model. Synergistic effects were noted of combined TNF/IL-1 blocking [31]. More recently, novel transgenic mouse models have been developed, which provide further insight into a role of IL-1 in generation of arthritogenic T cells. Mice deficient in IL-1Ra display uncontrolled IL-1 activity, and develop spontaneous T cell-dependent autoimmune arthritis in a defined genetic background [32]. The model is impaired in TNF-, IL-6- and IL-17-deficient mice [33, 34]. It argues that excessive IL-1, together with IL-6, generates autoreactive Th17 cells. When neutralizing antibodies are given after onset of arthritis, anti-TNF was ineffective, anti-IL-17 halted further progression, but anti-IL-1 reduced the arthritis (personal unpublished observations).

Other examples of manipulated T cell function leading to autoimmune arthritis are the SKG and GP 130 arthritis models [35–37]. In the SKG arthritis, aberrant T cell receptor function allows positive selection of autoimmune T cells, whereas in the GP 130 model a mutation in the IL-6 receptor induced enhanced signaling and identified excessive IL-6 signaling as being able to drive T cell-dependent autoimmune arthritis. SKG arthritis was strikingly impaired when the mice were crossed with IL-1-deficient mice.

Innate arthritis – Repeated exacerbations

Strongest TNF dependence of acute inflammation is found when arthritis is induced locally with a phlogistic trigger such as streptococcal cell wall fragments (SCW) or yeast particles (Zymosan). Both joint swelling and cell influx in the synovial tissue is markedly reduced with anti-TNF antibodies. A similar dependence of TNF is noted in TNF$^{-/-}$ mice. IL-1 has no role in this, but is responsible for the catabolic effect on the articular cartilage [38–40]. Optimal control of both inflammation and cartilage damage is achieved with anti-TNF/IL-1 combination therapy.

The pattern changes when repeat injections are given (Fig. 4), characterized by consecutive flares, more persistent synovial infiltrate and joint erosion. After three to four rechallenges, the swelling response remains a TNF-dependent phenomenon, but chronic cellular infiltration as well as bone and cartilage erosion become IL-1 dependent. In fact, marked synovitis and erosion is seen when this repeat model is induced in TNF-deficient mice, whereas this is absent in IL-1$^{-/-}$ mice [38, 40]. In addition, the model becomes IL-17 dependent. Apparently, arthritis starts with TLR2-mediated local activation of synovial macrophages and fibroblasts. After repeat challenges, T cell reactivity is generated in an IL-1-dependent fashion. IL-17 levels increase; the chronic model is markedly reduced in IL-17R-deficient mice [41].

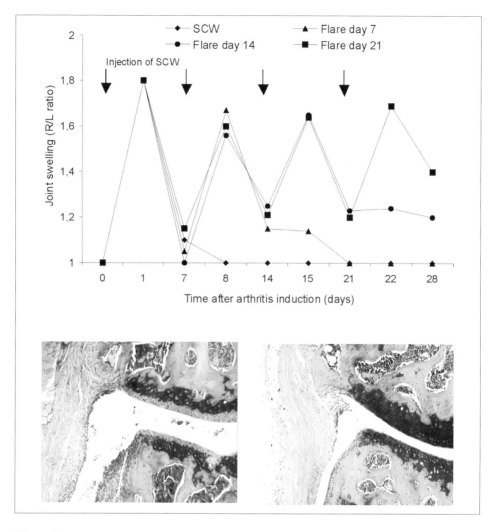

Figure 4
Pattern of arthritis induced by a single or consecutive repeat intra-articular injections of SCW fragments. Note the increasing chronicity. The histology shows a picture at day 7 after the fourth SCW rechallenge in control mice (left) and IL-1β-deficient mice (right). Inflammation and erosion is markedly reduced in the latter .

Of note, molecular mimicry has been demonstrated between SCW peptidogly-cans and cartilage proteoglycans at the T cell level and cross-reactive autoimmunity may contribute in the chronic phase. Cartilage fragments such as fibronectin, aggre-can and biglycan may contribute to arthritis *via* TLR4 activation [2].

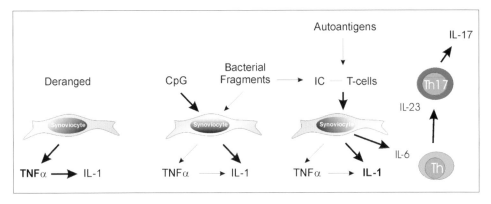

Figure 5
Cytokine patterns may change dependent on the driving elements. Shift in total TNF dependence of IL-1 production to an almost independent role, ultimately culminating in autonomous IL-17 production, where sufficient Th17 cells are generated with IL-1, IL-6 and IL-23 help.

Cytokine dependency may shift

Depending of the pathogenic pathway, relative cytokine dominance may shift (Fig. 5). Of course, when the underlying defect of arthritis is a deranged overproduction of TNF, then all IL-1 production is downstream of TNF and TNF blockade is sufficient. When synovial activation is caused by nonimmune macrophage/fibroblast activation, TNF still is a major player, although TNF-independent IL-1 production is evident. The pattern shifts to IL-1 and IL-17 when immune elements come into play. Immune complexes mainly drive TNF-independent IL-1 production, and IL-1 is a crucial factor in generation of autoreactive T cells [42]. Finally, the process may even become IL-1 independent when sufficient Th17 cells are generated, and direct triggering of such cells creates IL-17-driven arthritis. Such a condition is evident in pure T cell-mediated flares of AIA [27].

Potential overkill of IL-1 by other pathways

As mentioned above, IL-17 is a cytokine mainly derived from the recently identified T cell subset Th17. It shares many properties with IL-1, including its effect on articular chondrocytes to drive cartilage destruction and its potential to up-regulate RANKL and to mediate bone erosion. Although less potent than IL-1, it shows strong synergy with TNF and can greatly exaggerate arthritis driven by other stimuli, as identified in passive GPI arthritis [43]. Local IL-17 overexpression strongly exacerbates CIA, but also overrules the IL-1 dependency of this model [24].

Another intriguing finding was the capacity of the TLR4 agonist LPS to circumvent IL-1 dependency of passive KRN arthritis. When serum of arthritic KRN mice, containing anti-GPI antibodies, is passively transferred to normal recipients, these animals develop florid arthritis, which is absent in IL-1-deficient mice. However, when LPS is co-administered GPI arthritis develops undisturbed in IL-1$^{-/-}$ mice [44]. The TLR4 pathway shares many of the signaling elements of the IL-1 pathway and this might explain IL-1 redundancy. It argues that cytokine dependency of an arthritic process may shift when environmental TLR ligands such as bacteria and viruses become involved.

Anti-IL-1 therapy in the clinical situation

It has been long recognized that reduction of IL-1 is a powerful therapeutic approach to prevent chronic erosive joint inflammation in various murine arthritis models. If elements of the models apply to the clinical situation in RA patients, IL-1 directed therapy makes sense. There is growing evidence that autoantibodies contribute to severity and erosive character of RA. Efficacy of anti-B cell therapy and blocking of activated T cells (CTLA4) underline immune involvement in RA patients and would suggest that IL-1 is a crucial player. Nevertheless, clinical trials with IL-1Ra (anakinra) have been disappointing so far. Joint erosion was markedly suppressed but effects on joint inflammation were, at best, moderate. It has long been argued that IL-1Ra treatment might have been suboptimal, due to poor pharmacokinetics. This was also the impression of experimental arthritis work, where IL-1Ra continuously supplied at high dose with Alzet minipumps was efficacious in CIA, whereas daily dosing was insufficient [19]. However, the great efficacy of IL-1Ra treatment in adult-onset Still's disease [45], several autoinflammatory disorders linked to mutations of proteins controlling IL-1β secretion, and more recently gout, fueled the disbelief of a dominant role of IL-1 in RA. Novel anti-IL-1 therapies with high quality neutralizing anti-IL-1β antibodies [22] are under investigation at present, and will provide definite insight into the role of IL-1 in RA. It is warranted to pay proper attention to impact on cartilage erosion, which is a hallmark of IL-1 activity. Most trials just score joint space narrowing on X-rays, which is a poor read out of focal cartilage damage. Hopefully, improved MRI technology will become available to offer greater sensitivity.

References

1   van Lent PL, Grevers L, Lubberts E, de Vries TJ, Nabbe KC, Verbeek S, Oppers B, Sloetjes A, Blom AB, van den Berg WB (2006) FcgammaRs directly mediate cartilage

but not bone destruction in murine antigen induced arthritis: Uncoupling of cartilage damage from bone erosion and joint inflammation. *Arthritis Rheum* 54: 3868–77

2   Schaefer L, Babelova A, Kiss E, Hausse HJ, Baliova M, Kryzankova M, Marsche G, Young MF, Mihalik D, Gotte M et al (2005) The matrix component biglycan is proin-flammatory and signals through Toll-like receptors 4 and 2 in macrophages. *J Clin Invest* 115: 2223–2233

3   Arend WP, Dayer JM (1995) Inhibition of the production and effects of IL-1 and TNFα in RA. *Arthritis Rheum* 38: 151–160

4   Sims JE, Nicklin MJ, Bazan JF, Barton JL, Busfield SJ, Ford JE, Kastelein RA, Kumar S, Lin H, Mulero JJ et al (2001) A new nomenclature for IL-1-family genes. *Trends Immunol* 22: 536–537

5   Burger D, Dayer JM, Palmer G, Gabay C (2006) Is IL-1 a good therapeutic target in the treatment of arthritis? *Best Pract Res Clin Rheumatol* 20: 879–96

6   Dinarello CA (2005) Blocking IL-1 in systemic inflammation. *J Exp Med* 201: 1355–1359

7   Tschopp J, Martinon F, Burns K et al (2003) NALPs: A novel protein family involved in inflammation. *Nat Rev Mol Cell Biol* 4: 95–104

8   Elssner A, Duncan M, Gavrilin M, Wevers MD (2004) A novel P2X7 receptor activator, the human cathelicidin-derived peptide LL37, induces IL-1 beta processing and release. *J Immunol* 172: 4987–4994

9   Coeshott C, Ohnemus C, Pilyavskaya A, Ross S, Wieczorek M, Kroona H, Leimer AH, Chernis J (1999) Converting enzyme-independent release of TNFα and IL-1β from a stimulated human monocytic cell line in the presence of activated neutrophils or purified proteinase 3. *Proc Natl Acad Sci USA* 96: 6261–6266

10   Janssens S, Beyaert R (2003) Functional diversity and regulation of different interleu-kin-1 receptor associated kinases (IRAK) family members. *Mol Cell* 11: 293–302

11   Smeets RL, Joosten LA, Arntz OJ, Bennink MB, Takahashi N, Carlsen H, Martin MU, van den Berg WB, van de Loo FA (2005) Soluble IL-1RacP ameliorates collagen induced arthritis by a different mode of action from that of IL-1ra. *Arthritis Rheum* 52: 2202–11

12   Bresnihan B, Newmark R, Robbins S, Genant HK (2004) Effects of Anakinra mono-therapy on joint damage in patients with RA. Extension of a 24-week randomized, placebo-controlled trial. *J Rheumatol* 31: 1103–11

13   Van de Loo A.A.J. van den Berg WB (1990) Effects of murine recombinant IL-1 on synovial joint in mice: Measurement of patellar cartilage metabolism and joint inflam-mation. *Ann Rheum Dis* 49: 238–245

14   Niki Y, Yamada H, Seki S, Kikuchi T, Takaishi H, Toyama Y, Fujikawa K, Tada N (2001) Macrophage and neutrophil dominant arthritis in human IL-1 alpha transgenic mice *J Clin Invest* 107: 1127–35

15   Niki Y, Yamada H, Kikuchi T, Toyama Y, Matsumoto H, Fujikawa K, Tada N (2004) Membrane-associated IL-1 contributes to chronic synovitis and cartilage destruction in human IL-1alpha transgenic mice. *J Immunol* 172: 577–84

16 Probert L, Plows D, Kontogerogos G, Kollias G (1995) The type I IL-1 receptor acts in series with TNFα to induce arthritis in TNFα transgenic mice. *Eur J Immunol* 25: 1794–97

17 Zwerina J, Redlich K, Polzer K, Joosten LA, Kroenke G, Distler J, Hess A, Pundt N, Pap T, Hoffman O et al (2007) TNF-induced structural joint damage is mediated by IL-1. *Proc Natl Acad Sci USA* 104: 11742–47

18 Wei S, Kitaura H, Zhou P, Ross FP, Teitelbaum SL (2005) IL-1 mediates TNF-induced osteoclastogenesis. *J Clin Invest* 115: 282–290

19 Joosten LA, Helsen MA, van de Loo FA, van den Berg WB (1996) Anticytokine treatment of established type II collagen-induced arthritis in DBA/1 mice. A comparative study using anti-TNF alpha, anti-IL-1 alpha/beta, and IL-1Ra. *Arthritis Rheum* 39: 797–809

20 Mori L, Iselin S, Delibero G, Lesslauer W (1996) Attenuation of collagen-induced arthritis in 55-kD TNF receptor type I (TNFR1)-IgG1 treated and TNFR1-deficient mice. *J Immunol* 157: 3178–3182

21 Campbell IK, O'Donnel K, Lawlor KE, Wicks IP (2001) Severe inflammatory arthritis and lymphadenopathy on the absence of TNF. *J Clin Invest* 107: 1519–1527

22 Alten RH, Pohl CC, Batard YY, Wright AA, Gram HH, Bobadilla MM (2006) ACR 20/50/70 responses in MTX resistant RA patients in a double blind, placebo controlled phase 1/II evaluation of the pharmacokinetics / pharmacodynamics, safety, and preliminary efficacy of a fully human anti-IL-1β monoclonal antibody ACZ8985. *Ann Rheum Dis* 65 (Suppl II): 60

23 Ma YS, Thornton GP, Boivin D, Hirsch R, Hirsch E (1998) Altered susceptibility to collagen induced arthritis in transgenic mice with aberrant expression of IL-1ra. *Arthritis Rheum* 41: 1798–1805

24 Lubberts E, Koenders MI, van den Berg WB (2005) The role of IL-17 in conducting destructive arthritis: Lessons from animal models. *Arthritis Res Ther* 7: 29–37

25 van de Loo FAJ, Joosten LAB, van Lent PL, Arntz OJ, van den Berg WB (1995) Role of IL-1, TNF and IL-6 in cartilage proteoglycan metabolism and destruction: Effect of *in situ* cytokine blocking in murine antigen and zymosan induced arthritis. *Arthritis Rheum* 38: 164–72

26 van Meurs JBJ, van Lent PL, Singer II, Bayne EK, van de Loo FAJ, van den Berg WB (1998) IL-1ra prevents expression of the metalloproteinase generated neoepitope VDIPEN in antigen induced arthritis. *Arthritis Rheum* 41: 647–56

27 Koenders MI, Lubberst E, Oppers-Walgreen B, van den Bersselaar L, Helsen MMA, DiPadova FE, Boots AM, Gram H, Joosten LA, van den Berg WB (2005) Blocking of IL-17 during reactivation of experimental arthritis prevents joint inflammation and bone erosion by decreasing RANKL and IL-1. *Am J Pathol* 167: 141–49

28 van Lent PL, van de Loo FA, Holthuysen AE, van den Bersselaar LA, Vermeer H, van den Berg WB (1995) Major role for IL-1 but not for TNF in early cartilage damage in immune complex arthritis in mice. *J Rheumatol* 22: 2250–58

29   Monach PI, Benoist C, Mathis D (2004) The role of antibodies in mouse models of RA and relevance to human disease. *Adv Immunol* 82: 217–48

30   Nigrovic PA, Binstadt BA, Monach PA, Johnson A, Gurish M, Iwakura Y, Benoist C, Mathis D, Lee DM (2007) Mast cells contribute to initiation of autoantibody-mediated arthritis *via* IL-1. *Proc Natl Acad Sci USA* 104: 2325–30

31   Bendele AM, Chipala ES, Scherrer J, Frazier J, Sennelo G, Rich WJ, Edwards CK III (2000) Combination benefit of treatment with the cytokine inhibitors IL-1ra and PEGylated soluble TNF receptor type I in animals of rheumatoid arthritis. *Arthritis Rheum* 43: 2648–59

32   Horai R, Saijo S, Tanioka H, Nakae S, Sudo K, Okahara A, Ikuse T, Asano M, Iwakura Y (2000) Development of chronic inflammatory arthropathy resembling RA in IL-1ra deficient mice. *J Exp Med* 191: 313–20

33   Horai R, Nakajima A, Habiro K, Kotani M, Nakae S, Matsuki T, Nambu A, Saijo S, Kotaki H, Sudo K et al (2004) TNFα is crucial for the development of autoimmune arthritis in IL-1ra deficient mice. *J Clin Invest* 114: 1603–11

34   Nakae S, Saiijo S, Horai R, Sudo K, Mori S, Iwakura Y (2003) Il-17 production from activated T cells is required for the spontaneous development of destructive arthritis in mice deficient in IL-1ra. *Proc Natl Acad Sci USA* 100: 5986–90

35   Hata H, Sakaguchi N, Yoshitomi H, Iwakura Y, Sekikawa K, Azuma Y, Kanai C, Moriizumi E, Nomura T, Nakamura T et al (2004) Distinct contribution of IL-6, TNFα, IL-1 and IL-10 to T cell mediated spontaneous autoimmune arthritis in mice. *J Clin Invest* 114: 582–8

36   Hirota K, Hashimoto M, Yoshitomi H, Tanaka S, Nomura T, Yamaguchi T, Iwakura Y, Sakaguchi N, Sakaguchi S (2007) T cell self-reactivity forms a cytokine milieu for spontaneous development of IL-17 Th cells that cause autoimmune arthritis. *J Exp Med* 204: 41–7

37   Sawa S, Kamimura D, Jin GH, Morikawa H, Kamon H, Nishihara M, Ishihara K, Murakami M, Hirano T (2006) Autoimmune arthritis associated with mutated IL-6 receptor gp130 is driven by STAT3/IL-7 dependent homeostatic proliferation of CD4+ T cells. *J Exp Med* 203: 1459–70

38   van den Berg WB, Joosten LA, van de Loo FA (1999) TNFα and IL-1β are separate targets in chronic arthritis. *Clin Exp Rheumatol* 17 (Suppl 18) S105–114

39   van den Berg WB (2001) Anti-cytokine therapy in chronic destructive arthritis. *Arthritis Res Ther* 3: 18–26

40   Joosten LA, Abdollahi-Roodsaz S, Heuvelmans M, Helsen MM, van den Bersselaar LA, Oppers-Walgreen B, Koenders MI, van den Berg WB (2008) T cell dependency of chronic destructive arthritis induced by repeated local activation of TLR-driven pathways: Crucial role for both IL-1beta and IL-17. *Arthritis Rheum* 58: 98–108

41   Lubberts E, Schwarzenberger P, Huang W, Schurr JR, Peschon JJ, van den Berg WB, Kolls JK (2005) Requirement of IL-17 receptor signaling in radiation-resistant cells in the joint for full progression of destructive synovitis. *J Immunol* 175: 3360–68

42   O'Sullivan BJ, Thomas HE, Pai S, Santamaria P, Iwakura Y, Steptoe RJ, Kay TW,

Thomas R (2006) IL-1β breaks tolerance through expansion of CD25⁺ effector cells. *J Immunol* 176: 7278–87

43 Koenders MI, Lubberts E, van de Loo FA, Oppers-Walgreen B, van den Bersselaar L, Helsen MM, Joosten LA, van den Berg WB (2006) IL-17 acts independently of TNFα under arthritic conditions. *J Immunol* 176: 6262–69

44 Choe JY, Crain B, Wu SR, Corr M (2003) IL-1 receptor dependence of serum transferred arthritis can be circumvented by toll-like receptor-4 signaling. *J Exp Med* 197: 537–42

45 Kalliolas GD, Georgiou PE, Antonopoulos IA, Andonopoulos AP, Liossis SN (2007) Anakinra treatment in patients with adult-onset Still's disease is fast, effective , safe and steroid sparing: Experience from an uncontrolled trial. *Ann Rheum Dis* 66: 842–3

Interleukin-15

Jagtar Nijar Singh and Iain B. McInnes

Centre for Rheumatic Diseases, Division of Immunology, Infection and Inflammation, University of Glasgow, Glasgow, UK

Abstract

Interleukin-15 (IL-15) is a cytokine of the four-α-helix superfamily that mediates pleiotropic effects in regulating components of both the innate and adaptive immune system. It binds to a heterotrimeric receptor consisting of the common γ-chain receptor, IL-15/IL-2 receptor β-chain and unique IL-15 receptor α-chain. IL-15 is expressed at the mRNA level in a variety of cell lineages and is expressed as protein as part of the rapid early inflammatory response. It mediates activation of NK cells, T cells, neutrophils and macrophages and as such is considered a broad immune activating moiety. IL-15 expression has been described in a variety of inflammatory diseases, including particularly rheumatoid arthritis, psoriatic arthritis and reactive arthritis. Within synovial tissues in particular it has been ascribed an inflammatory role by virtue of its capacity to activate T cells, NK cells macrophages and neutrophils. Moreover, *in vivo* model studies suggest that IL-15 neutralisation leads to reduction in articular inflammation and damage. Early clinical trials have shown promise in that IL-15 blockade using a monoclonal antibody in rheumatoid arthritis patients lead to some trends to improvement, providing biological proof of concept.

Introduction

Cytokines such as TNF-α and IL-1 are established therapeutic targets with strong basic preclinical rationale taken through to clinical trials and clinical practice across a rage of disorders. IL-15 is a cytokine with structural similarities to IL-2 [1, 2] that has been implicated in both the innate and adaptive arms of the immune response. Furthermore, it is likely to be involved in autoimmune and inflammatory conditions and, therefore, is under evaluation for its therapeutic potential. This chapter reviews the biological structure of IL-15, summarise its expression in normal tissues and its implication in disease processes and finally reviews recent findings implicating IL-15 as a therapeutic target.

Structure and expression of IL-15

IL-15 is structurally similar to IL-2 and comprises of a four-α-helix structure. However, it has a much wider distribution and functional repertoire than IL-2 and, in particular, is generally pro-inflammatory distinct from the immune regulatory suppressive effects of IL-2 in the T cell compartment. IL-15 is expressed in many normal tissues and in a variety of cellular lineages including monocytes, dendritic cells (DCs) and fibroblasts [3, 4]. IL-15 mRNA expression is not correlated with protein detection evidence of significant post-translation regulatory control [5]. Generally, two IL-15 isoforms are generated: one with a long signalling peptide containing 48 amino acids that is generally secreted, and a shorter signalling peptide comprising of 21 amino acids that traffics to both the cytoplasm and nucleus [6–10]. Therefore, cell membrane expression may be important in extracellular interactions of IL-15, whereas cytoplasmic and nuclear pools may represent either an endogenous feedback system or a store that can be mobilised in the event of an acute challenge.

The IL-15 receptor is heterotrimeric and consists of a β-chain (shared with IL-2), the common γ-chain and a unique α-chain. The IL-15Rα is a type I transmembrane receptor and is structurally related to the IL-2R α-chain. The structure is interesting because it contains a sushi domain for cytokine binding as well as a long intracellular domain associated with a potential for cell signalling without receptor complexes. Furthermore, receptor splicing yields eight isoforms some of which are unable to bind IL-15 [3, 11, 12] and so may limit IL-15 expression by acting as negative feedback signals within the cell [13].

IL-15 signalling pathways

In most leucocytes (except perhaps mast cells) IL-15R signals through JAK1/3- and STAT3/5-dependent pathways [3, 4]. Further signals may be transduced through *src*-related tyrosine kinases and Ras/Raf/MAPK to fos/jun activation. Other implicated pathways include the Bcl-related proteins and may account for the role of IL-15 in apoptosis. In fibroblasts, IL-15Rα may act as a co-stimulator with other receptor superfamilies including the receptor tyrosine kinase Axl, which signals through PI3K, Akt and finally Bcl-2/Bcl-X$_L$ [14]. IL-15 and IL-15Rα may interact with the β/γ-chain on adjacent cells facilitating *trans* signalling [15]. This is due to IL-15 having two sites at which IL-15Rα can bind, therefore giving the possibility that one cytokine molecule could bind two receptors allowing the potential for bi-directional signalling [4]. This could be of particular importance in the expansion and control of CD8 T cell subsets and raises important issues for potential therapeutic targeting. The IL-15 molecule itself may be involved by reverse signalling whereby IL-15 is expressed as an integral membrane protein *via* the long signalling peptide. Ligation of such IL-15 leads to serine phosphorylation and activation of mitogen-activated protein kinases (ERK1/2 and p38). A further pathway involving small Rho-GTPase Rac3 may be activated in a similar fashion – both of these pathways may be implicated in monocyte activation [16, 17].

IL-15 bioactivity

IL-15 has many effects commensurate with broad receptor expression (Tab. 1). These effects, however, are often characterised on the basis of addition of exogenous IL-15 and confirmation of a role for endogenous molecule is rather sparse – this is important given the limited extracellular expression of the cytokine that is described in most systems.

Table 1 - Biological effects of IL-15

Cell type	Key effects	References
T lymphocytes	- activation / proliferation - cytokine production Th/c1 & Th/c2 - cytotoxicity - chemokinesis - cytoskeletal rearrangement - adhesion molecule expression - reduced apoptosis	
B lymphocyte	- Ig production - proliferation	
NK Cell	- cytotoxicity - cytokine production - reduced apoptosis - lineage development	
Macrophage	- dose dependent effect on activation - membrane expression – costimulation	
Osteoclast	- maturation - calcitonin receptor upregulation	[42]
Dendritic Cell	- maturation - activation	
Neutrophil	- activation - cytoskeletal rearrangement - cytokine release - reduced apoptosis	
Fibroblast	- reduced apoptosis	

T cells

T cells up-regulate IL-15Rα as a feature of early activation. IL-15 induces the proliferation of both CD4 and CD8 T cells thereby driving clonal expansion, both antigen specific and polyclonal. IL-2 release is induced and cytotoxicity may be enhanced in relevant cellular subsets [3, 18, 19]. Various membrane activation markers such as CD69 or FasL have been shown to be up-regulated [20, 21] mainly on CD45RO+ but not CD45RA+ cells [20]. IL-15 also promotes T cell chemokinesis and adhesion molecule redistribution [22–24]. IL-15 in turn up-regulates both chemokine expression and chemokine receptor levels to further enhance cellular migration and recruitment to tissues as required. IL-15 has generally been shown to favour development of type 1 responses. For example, synovial T cells are induced to release high concentrations of IFN-γ *in vitro* by IL-15, and T cells from HIV-infected patients produce more IFN-γ in the presence of high dose IL-15 [25]. However, other studies have shown that IL-15 induces IL-5 production from allergen-specific T cell clones implying evidence for a role in type 2 responses [26]. IL-15 is now established as having a critical role in maintenance of T cell memory in both the CD8 [27] and CD4 T cell compartments. In particular, studies in IL-15 transgenic mice infected with *Listeria monocytogenes* support a role for specific memory in the CD8+ compartment [28] with further reports suggesting the observations extending into the CD4+ T cells [29, 30].

Macrophages

IL-15 may act as an autocrine regulator of macrophages with low levels suppressing activity and high levels inducing both pro-inflammatory cytokine and chemokine production [31]. Human macrophages also constitutively express membrane-bound IL-15 and this may be of importance in their early activation. Both LPS and GM-CSF induce translocation of cytoplasmic stores of IL-15 to the cell surface where it is able to sustain T cell proliferation [32]. This activity may comprise a major pathway for IL-15 effector function in early innate responses.

Dendritic cells

IL-15 along with GM-CSF has been show to mature monocytes into DCs (CD1a+, DR+, CD14−). DCs could be further matured using LPS, TNF-α or CD40L into CD83+, DC-LAMP+ cells [33]. Some of these cells express Langerhans cell markers such as E-cadherin and CCR6. Further studies suggest that IL-15 is involved in promoting IL-12 and NO release from myeloid DC (mDC) and also perhaps IL-2 secretion [34, 35]. Moreover, IL-15 is broadly expressed in mDC and plasmacytoid

DC subsets contained within the synovial membrane of patients with rheumatoid and psoriatic arthritis (Lebre et al., in preparation). Thus, it is likely that IL-15 operates at an early stage to promote maturation and functional activation of DCs.

Neutrophils

Neutrophils express the IL-15 receptor and IL-15 can induce activation and rearrangement of the cytoskeleton [36, 37]. It also enhances phagocytosis, increases both mRNA transcription and translation of a variety of cytokines and chemokines as well as reducing apoptosis. The latter may be mediated by decreased levels of caspase 1 and 3 thereby modifying Bax expression [4, 38]. The functional importance of these has been confirmed *in vitro* using *Candida albicans* but the significance *in vivo* in autoimmune inflammatory and host defence states needs to be further evaluated.

Eosinophils and mast cells

IL-15 regulates eosinophil survival probably mediated through increased GM-CSF production leading to NF-κB nuclear translocation [39]. Mast cells have been known to respond to IL-15, especially *via* a novel IL-15 receptor in this lineage [40], although the existence of the latter is not yet confirmed to our knowledge. Nevertheless, there is significant evidence for IL-15 having an effector function in these cells, perhaps mediated *via* the conventional receptor system described above. IL-15 enhances proliferation of mast cells and delays their apoptosis, most probably through a Bcl-K_L pathway and local release of IL-4, when growth factors are withdrawn [41].

Fibroblasts

IL-15 is expressed on fibroblasts grown from a variety of tissues and may be important given the theoretical ability of these cells to modify the immune response. Membrane IL-15 on fibroblasts is thought to be able to activate both NK and T cells [43, 44] and may, through an Akt/PI3K pathway and Bcl-2, be able to sustain fibroblast survival [14].

IL-15 expression in inflammatory arthritis

The pleiotropic effects of IL-15 described earlier clearly render it a candidate cytokine in the pathogenesis of inflammatory arthritis. IL-15 mRNA and protein have been detected in rheumatoid arthritis (RA) synovial membrane by a number of investigators. IL-15 mRNA levels are present at higher levels in RA than in other disease-related synovial biopsies [45]. Although cautious interpretation of mRNA data is required, it is of interest that IL-15 mRNA levels are higher in patients prior to commencement of immune suppressive therapy. Concentrations of protein present are similar to levels of TNF-α or IL-12 detected in parallel assays [46], but are lower than other monokines, e.g., IL-6 and IL-18. IL-15 has also been measured in RA synovial fluid using soluble IL-15R α-chain in a novel receptor capture assay [46] in which IL-15 levels in RA synovial fluid correlate closely with those detected by ELISA. We have also detected IL-15 in synovial fluids derived from patients with psoriatic arthritis, suggesting that IL-15 may be present in a broad range of inflammatory arthropathies. Moreover, Raza and colleagues [47] examined patients with very early arthritis using synovial lavage and observed high levels of IL-15 expression in those that subsequently met criteria for RA, suggesting that this cytokine may have an important early role commensurate with its activities in innate immune function. Low levels of IL-15 are also present in sera of up to 40% of RA patients, although variable levels have been reported in distinct populations [48, 49]. Serum IL-15 expression does not correlate with disease subsets thus far recognised. Whereas RA serum TNF-α levels correlate with the presence of germinal centres in parallel synovial biopsies, IL-15 levels were elevated in patents in whom either germinal centres or diffuse lymphocytic infiltrative patterns were observed [16].

IL-15 expression in inflamed synovium is found in lining layer macrophages, together with synovial fibroblasts and endothelial cells [50–52]. Synovial T cells have also been reported to express membrane IL-15 [51]. The distribution of IL-15 is similar in psoriatic and reactive arthritis synovial membranes but expression is at reduced levels as compared to RA [51]. Of interest, both psoriatic and reactive arthritis synovium contain IL-2, with which IL-15 may exhibit counter-regulatory activities. IL-15 expression has also recently been detected in synovial membrane derived from juvenile RA patients [53], associated with IL-18, IL-12 and IFN-γ expression.

Factors that in turn drive synovial IL-15 expression are unclear. We have recently shown that activated T cells can induce IL-15 expression in macrophages *via* cognate interactions. Exposure of synovial fibroblasts to TNF-α or IL-1β also induces high levels of IL-15 expression, although we have rarely detected this in secreted form. Recent studies in dermal fibroblasts similarly demonstrated that TNF-α but not IFN-γ induces membrane expression of IL-15, which in turn can sustain T cell growth [54]. A further pathway promoting IL-15 production has been suggested in studies of synovial embryonic growth factor expression. Overexpression of the wingless (*Wnt*)5 and frizzled (*Fz*)5 ligand pair is associated with increased production and secretion of IL-15 by RA synovial fibroblasts, together with IL-6 and IL-8 [55]. Furthermore, suppression of *Wnt5* or *Fz5* using antisense, dominant-negative mutants or neutralising antibodies led to reduction in IL-15 expression [56]. Thus, a variety of stimuli including cellular feedback loops may promote IL-15 release in synovium.

Strategies to target IL-15 *in vivo*

The complexities of IL-15 physiology pose considerable difficulties in determining what should be the optimal therapeutic strategy. Thus far three protein-based approaches have been considered, namely use of: (i) neutralising antibodies directed against either IL-15 or its receptor subcomponents, (ii) soluble IL-15Rα and (iii) mutated IL-15 species, usually generated as fusion proteins. A further approach is to utilise small molecule signal pathway inhibitors aimed particularly at JAK/STAT pathways subserving IL-15 function. These are not yet specific to IL-15-mediated function but inhibit several common γ-chain receptor-mediated events. Several studies utilising these diverse approaches have been attempted, or are currently ongoing.

Inflammation model studies – IL-15 targeting *in vivo*

Several of the approaches mentioned above have been tested in relevant disease models. We have used full-length soluble IL-15Rα administration to manipulate IL-15 bioactivity *in vivo*. When sIL-15Rα is injected daily following antigen challenge the development of collagen induced arthritis (CIA) is suppressed, associated with delayed development of anti-collagen-specific antibodies (IgG2a) and with reduced antigen-specific IFN-γ and TNF-α production *in vitro* [57]. On discontinuation of sIL-15Rα administration, CIA developed to levels comparable with controls, suggesting that anti-inflammatory effects are transient. In subsequent studies we have generated targeted mutants of IL-15Rα and identified the sushi domain as essential for functional cytokine neutralisation [58]. Selected deletion of cysteine residues similarly disrupted folding to abrogate binding and function. Studies are ongoing to determine whether small molecule derivatives of sIL-15Rα are of therapeutic utility in the CIA model. This also provides opportunities to investigate the potential for dual targeting of synergistic cytokine activities, e.g. IL-15 and IL-18.

An alternate approach has been to generate mutant IL-15 forms that can specifically modify IL-15 activities. An IL-15/Fcγ2a fusion protein that antagonises the activities of IL-15 *in vitro* and lyses receptor-bearing cells, suppresses the onset of delayed-type hypersensitivity responses *in vivo*, associated with reduction in CD4[+] T cell infiltration [59]. This fusion protein has also proven effective *in vivo* in preventing rejection of murine islet cell allografts in combination with CTLA4/Fc [60]. Studies in CIA indicate that this fusion protein is effective in treating not only developing CIA but also established disease, and that after treatment disease recurrence is suppressed. This effect is associated with suppression of expression of a broad range of inflammatory cytokines [61]. Finally, anti-IL-15 antibody has been employed in informative studies *in vivo* using a psoriasis model in which human psoriatic biopsies engrafted to SCID mice have received human anti-IL-15 monoclonal antibody (AMG714) leading to rapid clearance of the psoriatic tissue pathogenesis [62].

Clinical studies targeting IL-15

Clinical studies in humans have been performed using the neutralising antibody, AMG714, and a monoclonal antibody targeting IL-2/15R β-chain, MIKβ2. The optimal approach in clinical trials has not yet been established. The fully human IgG1 monoclonal anti-IL-15 antibody AMG714 binds and neutralises the activity of soluble and membrane-bound IL-15 *in vitro*. AMG714 has been tested in two clinical trials in RA. In a 12-week, dose-ascending, placebo-controlled study, RA patients ($n = 30$) that had failed several previous DMARDs received a randomised, controlled, single dose of AMG714 (0.5–8 mg/kg) followed by open label weekly doses for 4 weeks. IL-15 neutralisation was well tolerated [63]. This study was not placebo-controlled throughout; however, encouraging signs of efficacy were obtained. Around 60% of patients achieved an ACR20 response with some 25% achieving an ACR70 improvement. In parallel studies, AMG714 was shown to inhibit endogenous RA synovial T cell activation and to suppress IL-15-induced cytokine release [63]. A subsequent dose-finding study has now been performed [64] in which RA patients received increasing fixed does (up to 280 mg per injection) of anti-IL-15 antibody every 2 weeks by subcutaneous injection for 3 months. An interim analysis indicated satisfactory tolerance compared with placebo and ACR20 improvements were observed in approximately 60% of recipients receiving higher doses of AMG714. No significant alterations in the levels of circulating leucocyte subsets, including NK cells and CD8+ memory T cells, were observed. Extension of this study was performed to compare the highest dose of AMG714 ($n = 121$) with placebo ($n = 58$). Significant improvements in ARC20 responses occurred in AMG714 recipients compared to placebo at weeks 12 and 16 of follow-up. Of note, however, ACR20 responses were not significantly different from placebo at week 14 (reflecting a higher placebo response at this time point), the pre-designated primary outcome time for this study. Clear and significant improvements in acute-phase reactants occurred in AMG714 recipients compared with placebo. Thus, although there is clear evidence of biological activity and biological proof of concept, larger confirmatory studies are now required to facilitate proper interpretation of these data and at this stage IL-15 should not as yet be considered a validated therapeutic target in RA.

Several outstanding issues remain in this clinical area. The relative role of IL-15 as a target compared with TNF and IL-6 is unclear. Its role in early T cell/DC interactions suggests that it may have some role in tolerance induction and therefore manipulation of IL-15 may offer potential in early disease beyond its capability in later RA, the only subjects thus far treated. IL-15 mediates effects on epithelial cells of the gut, keratinocytes, myocytes, hepatocytes and several CNS subsets indicating broad tissue effector function in host defence [65–70]. Elevated levels are detected in a variety of inflammatory diseases and there is momentum currently to explore its therapeutic role across a range of disorders. In particular psoriasis offers attractive potential based on expression patterns in disease tissue, the beneficial effects of IL-15 blockade in relevant models and the potential for interruption of IL-15 function in remitting-relapsing inflammatory disease typical in some psoriatic disease patterns. Finally, it will now be necessary to extend the range of modalities of blockade. Pre-clinical studies are underway using IL-15 mutant proteins and additional anti-IL-15 monoclonal antibodies are under consideration. A Phase I trial has been performed in which IL-15 was blocked using Mikβ1 monoclonal antibody in patients with large granular lymphocyte leukaemia [71] – this reagent is now being tested in a variety of inflammatory conditions. In particular, there is interest in utilising signal molecule inhibitors, e.g. JAK inhibitors, which are in ongoing clinical trials in RA with encouraging early results. These do not yet however facilitate specific cytokine targeting. This may not be a deficit in their strategic importance as focussing on a given pathological signalling pathways may offer some advantages over pan cytokine inhibition.

Conclusion

IL-15 and its receptor are expressed in a wide range of cell types. It contributes to a pathway involved in the early activation of the immune system and enhances NK, polymorphonuclear and T cell responses. It has been implicated in several of the inflammatory arthropathies and *in vivo* clinical trials suggest a role in attenuating the aberrant immune response. However, as ever, further trials are required with larger numbers of patients to further elucidate its effect, in particular the interaction with TNF-α and other pro-inflammatory cytokines.

References

1   Grabstein KH, Eisenman J, Shanebeck K, Rauch C et al (1994) Cloning of a T cell growth factor that interacts with the beta chain of the interleukin-2 receptor. *Science* 264: 965–968

2   Bamford R, Grant A, Burton J, Peters C et al (1994) The interleukin (IL) 2 receptor beta chain is shared by IL-2 and a cytokine, provisionally designated IL-T, that stimulates T-cell proliferation and the induction of lymphokine-activated killer cells. *Proc Natl Acad Sci USA* 91: 4940–4944

3   Waldmann TA, Tagaya Y (1999) The multifaceted regulation of interleukin-15 expression and the role of this cytokine in NK cell differentiation and host response to intracellular pathogens. *Annu Rev Immunol* 17: 19–49

4   Budagian V, Bulanova E, Paus R, Bulfone-Paus S (2006) IL-15/IL-15 receptor biology: A guided tour through an expanding universe. *Cytokine Growth Factor Rev* 17: 259–280

5   Bamford RN, DeFilippis AP, Azimi N, Kurys G et al (1998) The 5' untranslated region, signal peptide, and the coding sequence of the carboxyl terminus of IL-15 participate in its multifaceted translational control. *J Immunol* 160: 4418–4426

6   Tagaya Y, Kurys G, Thies TA, Losi JM et al (1997) Generation of secretable and non-secretable interleukin-15 isoforms through alternate usage of signal peptides. *Proc Natl Acad Sci USA* 94: 14444–14449

7   Nishimura H, Washizu J, Nakamura N, Enomoto A et al (1998) Translational Efficiency Is up-regulated by alternative exon in murine IL-15 mRNA. *J Immunol* 160: 936–942

8   Gaggero A, Azzarone B, Andrei C, Mishal Z et al (1999) Differential intracellular trafficking, secretion and endosomal localization of two IL-15 isoforms. *Eur J Immunol* 29: 1265–1274

9   Nishimura H, Fujimoto A, Tamura N, Yajima T et al (2005) A novel autoregulatory mechanism for transcriptional activation of the IL-15 gene by a nonsecretable isoform of IL-15 generated by alternative splicing. *FASEB J* 19: 19–28

10   Dubois S, Magrangeas F, Lehours P, Raher S et al (1999) Natural splicing of exon 2 of human interleukin-15 receptor α chain mRNA results in shortened form with a distinct pattern of expression. *J Biol Chem* 274: 26978–84

11   Waldmann T, Tagaya Y, Bamford R (1998) Interleukin-2, interleukin-15, and their receptors. *Int Rev Immunol* 16: 205–226

12   Tagaya Y, Bamford RN, DeFilippis AP, Waldmann TA (1996) IL-15: A pleiotropic cytokine with diverse receptor/signaling pathways whose expression is controlled at multiple levels. *Immunity* 4: 329–336

13   Dubois S, Magrangeas F, Lehours P, Raher S et al (1999) Natural splicing of exon 2 of human interleukin-15 receptor α chain mRNA results in shortened form with a distinct pattern of expression. *J Biol Chem* 274: 26978–26984

14   Budagian V, Bulanova E, Orinska Z, Thon L et al (2005) A promiscuous liaison between IL-15 receptor and Axl receptor tyrosine kinase in cell death control. *EMBO J* 24: 4260–70

15   Dubois S, Mariner J, Waldmann TA, Tagaya Y (2002) IL-15Ra recycles and presents IL-15 in trans to neighboring cells. *Immunity* 17: 537–47

16   Budagian V, Bulanova E, Orinska Z, Pohl T et al (2004) Reverse signaling through membrane-bound interleukin-15. *J Biol Chem* 279: 42192–201

17   Neely GG, Epelman S, Ma LL, Colarusso P, Howlett CJ, Amankwah EK et al (2004) Monocytes surface-bound IL-15 can function as an activating receptor and participate in reverse signaling. *J Immunol* 172: 4225–34

18   Nishimura H, Hiromatsu K, Kobayashi N, Grabstein KH et al (1996) IL-15 is a novel

growth factor for murine gamma delta T cells induced by *Salmonella* infection. *J Immunol* 156: 663–669

19   Korholz D, Banning U, Bonig H, Grewe M et al (1997) The role of interleukin-10 (IL-10) in IL-15–mediated T-cell responses. *Blood* 90: 4513–4521

20   Kanegane H, Tosato G (1996) Activation of naive and memory T cells by interleukin-15. *Blood* 88: 230–235

21   Mottonen M, Isomaki P, Luukkainen R, Toivanen P et al (2000) Interleukin-15 up-regulates the expression of CD154 on synovial fluid T cells. *Immunology* 100: 238–244

22   Wilkinson PC, Liew FY (1995) Chemoattraction of human blood T lymphocytes by interleukin-15. *J Exp Med* 181: 1255–1259

23   Al-Mughales J, Blyth TH, Hunter JA, Wilkinson PC (1996) The chemoattractant activity of rheumatoid synovial fluid for human lymphocytes is due to multiple cytokines. *Clin Exp Immunol* 106: 230–236

24   Nieto M, del Pozo MA, Sanchez-Madrid F (1996) Interleukin-15 induces adhesion receptor redistribution in T lymphocytes. *Eur J Immunol* 26: 1302–1307

25   Seder RA, Grabstein KH, Berzofsky JA, McDyer JF (1995) Cytokine interactions in human immunodeficiency virus-infected individuals: Roles of interleukin (IL)-2, IL-12, and IL-15. *J Exp Med* 182: 1067–1077

26   Mori A, Suko M, Kaminuma O, Inoue S et al (1996) IL-15 promotes cytokine production of human T helper cells. *J Immunol* 156: 2400–2405

27   Tough DF, Zhang X, Sprent J (2001) An IFN-gamma-dependent pathway controls stimulation of memory phenotype CD8+ T cell turnover *in vivo* by IL-12, IL-18, and IFN-gamma. *J Immunol* 166: 6007–6011

28   Yajima T, Nishimura H, Ishimitsu R, Watase T et al (2002) Overexpression of IL-15 *in vivo* increases antigen-driven memory CD8+ T cells following a microbe exposure. *J Immunol* 168: 1198–1203

29   Geginat J, Sallusto F, Lanzavecchia A (2001) Cytokine-driven proliferation and differentiation of human naive, central memory, and effector memory CD4+ T cells. *J Exp Med* 194: 1711–1720

30   Niedbala W, Wei X, Liew FY (2002) IL-15 induces type 1 and type 2 CD4+ and CD8+ T cells proliferation but is unable to drive cytokine production in the absence of TCR activation or IL-12/IL-4 stimulation *in vitro*. *Eur J Immunol* 32: 341–347

31   Alleva DG, Kaser SB, Monroy MA, Fenton MJ et al (1997) IL-15 functions as a potent autocrine regulator of macrophage proinflammatory cytokine production: Evidence for differential receptor subunit utilization associated with stimulation or inhibition. *J Immunol* 159: 2941–2951

32   Neely GG, Robbins SM, Amankwah EK, Epelman S et al (2001) Lipopolysaccharide-stimulated or granulocyte-macrophage colony-stimulating factor-stimulated monocytes rapidly express biologically active IL-15 on their cell surface independent of new protein synthesis. *J Immunol* 167: 5011–5017

33   Mohamadzadeh M, Berard F, Essert G, Chalouni C et al (2001) Interleukin 15 skews

monocyte differentiation into dendritic cells with features of Langerhans cells. *J Exp Med* 194: 1013–1020

34   Ohteki T, Suzue K, Maki C, Ota T et al (2001) Critical role of IL-15–IL-15R for antigen-presenting cell functions in the innate immune response. *Nat Immunol* 2: 1138–1143

35   Feau S, Facchinetti V, Granucci F, Citterio S et al (2005) Dendritic cell-derived IL-2 production is regulated by IL-15 in humans and in mice. *Blood* 105: 697–702

36   Girard D, Paquet ME, Paquin R, Beaulieu AD (1996) Differential effects of interleukin-15 (IL-15) and IL-2 on human neutrophils: Modulation of phagocytosis, cytoskeleton rearrangement, gene expression, and apoptosis by IL-15. *Blood* 88: 3176–3184

37   Girard D, Boiani N, Beaulieu AD (1998) Human neutrophils express the interleukin-15 receptor alpha chain (IL-15Ralpha) but not the IL-9Ralpha component. *Clin Immunol Immunopathol* 88: 232–240

38   Bouchard A, Ratthe C, Girard D (2004) Interleukin-15 delays human neutrophils apoptosis by intracellular events and not *via* extracellular factors: Role of Mcl-1 and decreased activity of caspase-3 and caspase-8. *J Leukoc Biol* 75: 893–900

39   Hoontrakoon R, Chu WH, Gardai SJ, Wenzel SE et al (2002) Interleukin-15 inhibits spontaneous apoptosis in human eosinophils *via* autocrine production of granulocyte macrophage-colony stimulation factor and nuclear factor-κB activation. *Am J Respir Cell Mol Biol* 26: 404–12

40   Tagaya Y, Bamford RN, DeFilippis AP, Waldmann TA (1996) IL-15: A pleiotropic cytokine with diverse receptor/signaling pathways whose expression is controlled at multiple levels. *Immunity.* 4: 329–36

41   Masuda A, Matsuguchi T, Yamaki K, Hayakawa T, Yoshikai Y (2001) Interleukin-15 prevents mouse mast cell apoptosis through STAT6–mediated Bcl-xL expression. *J Biol Chem* 276: 26107–13

42   Ogata Y, Kukita A, Kukita T, Komine M et al (1999) A novel role of IL-15 in the development of osteoclasts: Inability to replace its activity with IL-2. *J Immunol* 162: 2754–2760

43   Briard D, Brouty-Boye D, Azzarone B, Jasmin C (2002) Fibroblasts from human spleen regulate NK cell differentiation from blood CD34$^+$ progenitors *via* cell surface IL-15. *J Immunol* 168: 4326–32

44   Rappl G, Kapsokefalou A, Heuser C, Rossler M et al (2001) Dermal fibroblasts sustain proliferation of activated T cells *via* membrane-bound interleukin-15 upon long-term stimulation with tumor necrosis factor-α. *J Invest Dermatol* 116: 102–9

45   Kotake S, Schumacher HR Jr, Yarboro CH, Arayssi TK et al (1997) *In vivo* gene expression of type 1 and type 2 cytokines in synovial tissues from patients in early stages of rheumatoid, reactive, and undifferentiated arthritis. *Proc Assoc Am Physicians* 109: 286–301

46   McInnes IB, Leung BP, Feng GJ, Sturrock RD et al (1998) A role for IL-15 in rheumatoid arthritis. *Nat Med* 4: 645

47   Raza K, Falciani F, Curnow SJ, Ross EJ et al (2005) Early rheumatoid arthritis is char-

acterized by a distinct and transient synovial fluid cytokine profile of T cell and stromal cell origin. *Arthritis Res Ther* 7: R784–95

48   Aringer M, Stummvoll GH, Steiner G, Koller M et al (2001) Serum interleukin-15 is elevated in systemic lupus erythematosus. *Rheumatology* 40: 876–881

49   Klimiuk PA, Sierakowski S, Latosiewicz R, Cylwik B et al (2001) Serum cytokines in different histological variants of rheumatoid arthritis. *J Rheumatol* 28: 1211–1217

50   McInnes IB, al-Mughales J, Field M, Leung BP et al (1996) The role of interleukin-15 in T-cell migration and activation in rheumatoid arthritis. *Nat Med* 2: 175–182

51   Thurkow EW, van der Heijden IM, Breedveld FC, Smeets TJ et al (1997) Increased expression of IL-15 in the synovium of patients with rheumatoid arthritis compared with patients with *Yersinia*-induced arthritis and osteoarthritis. *J Pathol* 181: 444–450

52   Oppenheimer-Marks N, Brezinschek RI, Mohamadzadeh M, Vita R et al (1998) Interleukin 15 is produced by endothelial cells and increases the transendothelial migration of T cells *in vitro* and in the SCID mouse-human rheumatoid arthritis model *in vivo*. *J Clin Invest* 101: 1261–1272

53   Scola MP, Thompson SD, Brunner HI, Tsoras MK et al (2002) Interferon-gamma: Interleukin 4 ratios and associated type 1 cytokine expression in juvenile rheumatoid arthritis synovial tissue. *J Rheumatol* 29: 369–378

54   Rappl G, Kapsokefalou A, Heuser C, Rossler M et al (2001) Dermal fibroblasts sustain proliferation of activated T cells *via* membrane-bound interleukin-15 upon long-term stimulation with tumor necrosis factor-alpha. *J Invest Dermatol* 116: 102–109

55   Sen M, Lauterbach K, El-Gabalawy H, Firestein GS et al (2000) Expression and function of wingless and frizzled homologs in rheumatoid arthritis. *Proc Natl Acad Sci USA* 97: 2791–2796

56   Sen M, Chamorro M, Reifert J, Corr M et al (2001) Blockade of Wnt-5A/frizzled 5 signaling inhibits rheumatoid synoviocyte activation. *Arthritis Rheum* 44: 772–781

57   Ruchatz H, Leung BP, Wei XQ, McInnes IB et al (1998) Soluble IL-15 receptor alpha-chain administration prevents murine collagen-induced arthritis: A role for IL-15 in development of antigen-induced immunopathology. *J Immunol* 160: 5654–5660

58   Wei X, Orchardson M, Gracie JA, Leung BP et al (2001) The Sushi domain of soluble IL-15 receptor alpha is essential for binding IL-15 and inhibiting inflammatory and allogenic responses *in vitro* and *in vivo*. *J Immunol* 167: 277–282

59   Kim YS, Maslinski W, Zheng XX, Stevens AC et al (1998) Targeting the IL-15 receptor with an antagonist IL-15 mutant/Fc gamma2a protein blocks delayed-type hypersensitivity. *J Immunol* 160: 5742–5748

60   Ferrari-Lacraz S, Zheng XX, Kim YS, Li Y et al (2001) An antagonist IL-15/Fc protein prevents costimulation blockade-resistant rejection. *J Immunol* 167: 3478–3485

61   Ferrari-Lacraz S, Zanelli E, Neuberg M, Donskoy E et al (2004) Targeting IL-15 receptor-bearing cells with an antagonist mutant IL-15/Fc protein prevents disease development and progression in murine collagen-induced arthritis. *J Immunol* 173: 5818–26

62 Villadsen LS, Schuurman J, Beurskens F, Dam TN et al (2003) Resolution of psoriasis upon blockade of IL-15 biological activity in axenograft mouse model. *J Clin Invest* 112: 1571–80

63 Baslund B, Tvede N, Danneskiold-Samsoe B, Larsson P et al (2005) Targeting interleukin-15 in patients with rheumatoid arthritis: A proof-of-concept study. *Arthritis Rheum* 52: 2686–92

64 Mcinnes IB, Martin R, Zimmermann-Gorska I, Nayiager S et al (2006) Safety and efficacy of a human monoclonal antibody to IL-15 (AMG 714) in patients with rheumatoid arthritis: Results of a multicenter, randomized, double-blind, placebo-controlled trial *Ann Rheum Dis* 65: EULAR Supplement

65 Kirman I, Nielsen OH (1996) Increased numbers of interleukin-15-expressing cells in active ulcerative colitis. *Am J Gastroenterol* 91: 1789–1794

66 Kakumu S, Okumura A, Ishikawa T, Yano M et al (1997) Serum levels of IL-10, IL-15 and soluble tumour necrosis factor-alpha (TNF-alpha) receptors in type C chronic liver disease. *Clin Exp Immunol* 109: 458–463

67 Kivisakk P, Matusevicius D, He B, Soderstrom M et al (1998) IL-15 mRNA expression is up-regulated in blood and cerebrospinal fluid mononuclear cells in multiple sclerosis (MS). *Clin Exp Immunol* 111: 193–197

68 Stegall T, Krolick KA (2001) Myocytes respond *in vivo* to an antibody reactive with the acetylcholine receptor by upregulating interleukin-15: An interferon-gamma activator with the potential to influence the severity and course of experimental myasthenia gravis. *J Neuroimmunol* 119: 377–386

69 Agostini C, Trentin L, Facco M, Sancetta R et al (1996) Role of IL-15, IL-2, and their receptors in the development of T cell alveolitis in pulmonary sarcoidosis. *J Immunol* 157: 910–918

70 Muro S, Taha R, Tsicopoulos A, Olivenstein R et al (2001) Expression of IL-15 in inflammatory pulmonary diseases. *J Allergy Clin Immunol* 108: 970–975

71 Morris JC, Janik JE, White JD, Fleisher TA et al (2006) Preclinical and phase I clinical trial of blockade of IL-15 using Mikbeta1 monoclonal antibody in T cell large granular lymphocyte leukemia. *Proc Natl Acad Sci USA* 103: 401–6

IL-17 and Th17 cells, key players in arthritis

Pierre Miossec, Ling Toh and Saloua Zrioual

Department of Immunology and Rheumatology, Hôpital Edouard Herriot,
69437 Lyon Cedex 03, France

Abstract

IL-17 was identified in 1995/96 as a T cell-derived cytokine with effects on inflammation and neutrophil activation. Rheumatoid arthritis (RA) has emerged as the best-studied situation to justify the selection of IL-17 as a therapeutic target. By interacting with other proinflammatory cytokines, IL-17 was found to induce bone and cartilage destruction. In 2006, the precise cell source of IL-17 was identified in the mouse. These cells were named Th17, and a key role for these cells was demonstrated in various situations associated with inflammation. These new findings confirmed and extended the results previously obtained following the identification of IL-17 as a T cell-derived cytokine. At the same time, additional information was obtained on the other members of the IL-17 family and on the structure of the IL-17 receptor complex. Such knowledge has further extended the choice of possible modalities to control IL-17.

Introduction

Interest in IL-17 increased further recently, when in 2006, the precise cell source of IL-17 was identified in the mouse. These cells were named Th17, and a key role for these cells was demonstrated in various situations associated with inflammation. These new findings confirmed and extended the results previously obtained following the identification of IL-17 as a T cell-derived cytokine.

Demonstration of the role of IL-17 in many inflammatory conditions further supported the concept of IL-17 targeting for treatment. We review these new findings in light of the previous knowledge [1]. We focus on rheumatoid arthritis (RA), which has emerged as the best studied situation to justify the selection of IL-17 as a therapeutic target.

Identification of IL-17

IL-17 was described in 1995/96 as a proinflammatory cytokine produced by T cells. The key experiment was the demonstration that addition of IL-17 to mesenchymal

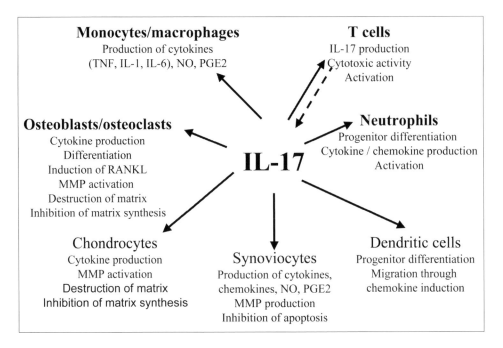

Figure 1
Effects of IL-17 on interactions between cells and cytokines associated with inflammation, cartilage, and bone destruction in rheumatoid arthritis.

cells/fibroblasts was able to increase IL-6 and other proinflammatory cytokine production, indicating immediately its role in inflammation [2, 3]. At the same time, IL-17 was shown to be able to induce neutrophil induction and maturation, an indication of its role in the acute mechanisms in host defense (Fig. 1).

Following the discovery of the molecule and of its key properties, a number of both human and mouse, spontaneous as well as induced, diseases were shown to be associated with IL-17 overexpression and production. For instance, IL-17 was on the list of genes, obtained through extensive gene array studies, found to be overexpressed in the brain of patients with multiple sclerosis. Using different approaches, similar conclusions were reached for Crohn's disease and psoriasis. Although the list of diseases will certainly increase, RA emerged as the best-studied situation making IL-17 a therapeutic target. Although we focus here mainly on this condition, it should be kept in mind that most of the results can probably be transferred to the long list of conditions where chronic inflammation is associated with matrix destruction, with the examples of myelin or bone and cartilage, respectively.

Concordant results showed that IL-17 was involved in the proinflammatory patterns associated with joint inflammation and RA using mouse and human models (Fig. 1) [4]. In the mouse, an injection of IL-17 alone into a normal knee was sufficient to induce cartilage damage [5]. The continuous administration of IL-17 by gene overexpression induced massive damage with extensive inflammatory cell migration, bone erosions, and cartilage degradation [6]. Conversely, inhibition with specific inhibitors including blocking antibodies and soluble receptor, or with IL-4 acting more broadly on other cytokines, provided protection from inflammation and destruction [7]. The studies showed an increased production of functional IL-17 by RA synovium but also by bone explants, indicating the role of T cells in juxta-articular bone destruction [8, 9]. As expected, this effect was associated with RANK ligand expression by these T cells, interacting with RANK-expressing cells, certainly osteoclasts but also mature dendritic cells [9, 10]. To clarify the role of IL-17 in chronicity, it was recently shown that in the mouse collagen arthritis model, IL-17 effect was dependent on the presence of TNF at the early phase, whereas at a later stage the disease was mostly IL-17 driven, and no longer TNF dependent [11].

Interactions between IL-17 and TNF

TNF is now considered as the key cytokine in RA. To consider the role of IL-17, it was necessary, to some extent, to take into account the role of T cells, or at least of some T cells. A key property for IL-17 was the synergistic interactions with TNF and to a lesser extent with IL-1 [12]. This critical property indicated that T cells could directly contribute to the inflammatory response. Furthermore, IL-17 increased the production of TNF and IL-1 by monocytes [13]. Opposite results were observed with IL-4, the prototype of a Th2 cytokine. This was another indication that IL-17-producing cells were a particular subset of T cells.

The sequential analysis indicated that synergy was observed only when cells such as synoviocytes were first exposed to IL-17 and then to TNF. Synergy was not observed when cells were exposed first to TNF then to IL-17. This is in line with the role of these Th17 cells in the amplification of the initial response associated with TNF and IL-1 secretion.

Th17, the cellular source of IL-17

The most critical and recent step was the identification in 2006 of the cell source of IL-17 [14]. IL-17 was first described as a T cell product [2]. IFN-γ is characteristic of Th1 cells and IL-4 of Th2 cells. The source of IL-17 was found to be different and these cells were named ThIL-17 or Th17 for short (Fig. 2). In the mouse, this new subset was identified by the demonstration of the inhibitory effect of IFN-γ on IL-17 production in mouse models of autoimmune diseases [15–17]. Induction of IFN-γ

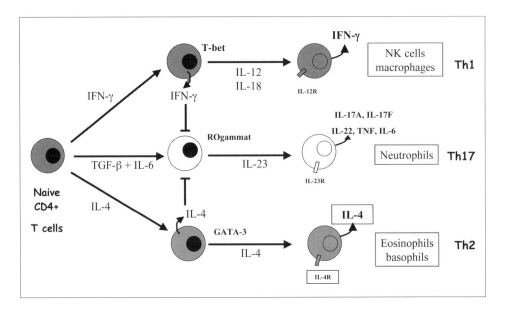

Figure 2
Th17: A new subset of Th cells.

was shown to be sensitive to the synergistic interaction between IL-12 and IL-18, two classical monocyte products [18]. The new finding was the inhibition of IL-17 by IFN-γ. The next step was the discovery of IL-23, another monocyte product, shown to be a key cytokine in the induction of brain inflammation in experimental models of encephalomyelitis [19].

IL-12 and IL-23 are two cytokines of the same family, both composed of dimers [20]. IL-12 is a heterodimer composed of the p35 and p40 subunits, whereas IL-23 is composed of the p19 and p40 subunits. The first antibodies against IL-12 were in fact against the p40 protein, which is common to IL-12 and IL-23. The protective effects first thought to be due to the inhibition of IL-12 through p40, could also result from IL-23 inhibition. This was shown by the specific inhibition of p19, the IL-23-specific subunit, whereas no effect was shown with the inhibition of p35 [21]. Conclusive demonstration came from the use of mice deficient for these IL-12- and IL-23-specific subunits. Finally, the enhancing effect of IL-23 on IL-17 production was demonstrated in various models of autoimmune diseases. We do not know yet how these results on IL-23 can be applied to human RA.

Transcription factors have been identified as markers of the Th subsets with T-bet for Th1, and GATA-3 for Th2 cells. The transcription factor retinoic acid-related orphan receptor γ t (RORγt) was found to be associated with Th17. Indeed, mice with RORγt-deficient T cells have attenuated autoimmune disease, and do not have tissue-infiltrating Th17 cells [22]. Conversely, T-bet inhibits IL-17 production *in vivo* [23].

Cytokine receptors have been employed the same way using the IL-12-specific IL-12R β2 chain, as a marker of Th1 cells, and the IL-23-specific IL-23R-specific chain, as a marker of Th17 cells. The IL-12R β1 chain is common to both the IL-12 and IL-23 receptors. The inhibitory effect of IL-17 on the Th1 pathway results from the IL-17 induced inhibition of IL-12R β2 expression, making cells not responsive to IL-12 effects on IFN-γ production.

IL-17 can induce IL-1 and TNF production by monocytes [13]. This effect on the amplification of inflammation was further expanded by showing that IL-6 and IL-1, two key proinflammatory cytokines, could in turn increase IL-17 production through the induction of IL-23, leading then to an increase in IL-17 production [24].

These results have been obtained in the mouse and as such should be considered with caution when applied to the human situation. Even in the mouse, more recent results indicated the frequent co-expression of IFN-γ and IL-17. Contribution of one of the two cytokines could lead to different pathogenic pathways, leading to the same clinical presentation as observed in mouse models of autoimmunity [25, 26]. Our results with RA T cells clones indicated that IL-17 was often produced in association with IFN-γ but not with IL-4 [27]. *In situ* immunostaining of the RA synovium showed two isolated populations of T cells producing either IFN-γ or IL-17. Double-positive cells were rarely seen. It thus remains to be demonstrated whether these secreting patterns are still dynamic or fixed. Of interest in this context was the demonstration that cytokine-secreting T cells have a particular morphology with a plasma cell appearance, a pattern that can be induced *in vitro* and is associated with the loss of TCR and CD3 but not of CD4 [28]. As for B cells, the plasma cell morphology of the IFN-γ and IL-17-producing cells strongly suggests that this is a fixed pattern related to a final stage of differentiation.

Th17/regulatory T cell balance

A key issue is the interaction between Th17 and regulatory T cells. Regulatory T cells are in charge of the control of the immune response. At baseline, in the absence of any particular stress or aggression, regulatory T cells are active and limit the intensity of the baseline response. In the context of stress, as seen during infection, this control is turned off to let the defense mechanisms be expressed according to the stimulation.

The first line of defense following exogenous stress results from the early nonspecific stimulation of monocytes, followed by a second wave of inflammatory signals from T cells. The related inflammation through the production of proinflammatory cytokines such IL-6 and IL-1 activates the Th17 pathway. At the same time the regulatory T cell pathway is inhibited. TGF-β alone is a potent inducer of regulatory T cells, whereas when combined with IL-6, an opposite effect is observed, leading to Th17 activation. TGF-β has been known for a long time for its immunosuppressive effects on lymphocytes. This new effect with enhancement of IL-17 production was thus more surprising [29]. In fact, it results more specifically from its inhibitory effects on the Th1 and Th2 pathways.

In the context of a chronic inflammatory disease such as RA, this situation is maintained, leading to the induction of disease-associated mechanisms and the inhibition of protective mechanisms. Although regulatory T cells can be seen at the site of inflammation, their function is defective resulting in chronicity. Induction of functional regulatory T cells has been observed in patients responding to TNF inhibition. Such a mechanism may well be critical in explaining the mode of action of TNF inhibition.

Recent papers indicated that TGF-β and IL-6 have much more potent effects on IL-17 production than IL-23 [30]. Regarding the regulatory balance between Th17 cells and regulatory T cells, IL-6 and TGF-β induce Th17 cells, whereas IL-6 inhibits the generation of Foxp3-positive regulatory T cells induced by TGF-β [30]. Conversely, the inhibition of IL-17 by Th2 products, i.e., IL-4, established a long time ago, was confirmed by showing the inhibition of TGF-β production by IL-4 [29, 31]. The human situation appears different where the combination of TGF-β and IL-21, and not IL-6, was shown to induce Th17 development and IL-1 could be more critical than IL-6 [32, 33].

The IL-17 family

The published results mainly refer to IL-17A, the founding member of the IL-17 family, which includes IL-17A–F. IL-17F has been given more attention because of its 50% sequence homology with IL-17A. When IL-17F is used alone, it appears to have similar effects to IL-17A but to a lower extent [34]. Sometimes IL-17F has minimal or even no effect when used alone. However, when combined with TNF, a synergistic effect is observed, almost as potent as with IL-17A. In contrast to the proinflammatory effects of IL-17A and F, IL-17E (also named IL-25) acts as a Th2 cytokine with anti-inflammatory properties [35, 36].

Th17 cells can produce several proinflammatory cytokines including IL-17A, IL-17F, IL-22, TNF, and IL-6. IL-22 is a member of the IL-10 family, and synergizes with IL-17A or IL-17F to regulate genes associated with skin innate immunity [37]. Recently, IL-22 was shown to mediate IL-23-induced dermal inflammation [38]. In addition, Th17 cells are involved in cell interactions through the expression of RANKL. Such RANK-RANKL interaction is the final bridge whereby osteoblasts activate osteoclasts leading to bone destruction. Similar interactions are found between synoviocytes and dendritic cells.

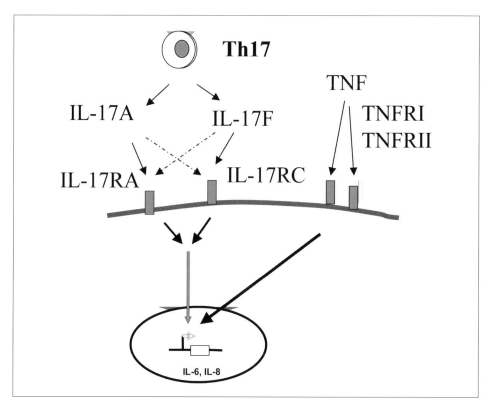

Figure 3
IL-17 A and IL-17F interactions with IL-17 receptors.

Structure of the IL-17 receptor

Not only the IL-17 family but also the IL-17 receptor (IL-17R) family have been characterized (Fig. 3). Some limited knowledge on the IL-17R was apparent from the first results demonstrating a rather low affinity for the IL-17R, suggesting the presence of additional chains [39, 40]. Sequence screening showed proteins with a partial homology with the IL-17R. In the mouse, at least two members have to be taken into account [41]. The first is the original IL-17R renamed IL-17RA. The second is IL-17RC. The physical association of the two receptors has been shown, although it is still unclear if these are two chains of a single receptor or two different receptors. It was previously proposed that IL-17A could be the receptor for IL-17A and IL-17RC the receptor for IL-17F. In the human situation, our impression is that the two receptors can bind either IL-17A or IL-17F, possibly with different affinities. These findings result from small interference RNA studies, leading to the inhibition of the cell surface expression of one of the two receptors. Inhibition of expression of one of the two receptors is sufficient to inhibit response to IL-17 alone in synoviocytes. However, the inhibition of the two receptors is needed to reduce the response to the combination of IL-17 with TNF [34]. As indicated previously, the effect of IL-17 used first increases that of TNF used in a second phase.

IL-17 targeting for treatment

The efficacy of cytokine targeting has first shown with TNF inhibitors. Similarly, the first available tools to inhibit IL-17 were a mouse soluble IL-17R receptor and an anti-IL-17 mouse monoclonal antibody. These tools were the equivalent of Enbrel and Remicade, developed to block TNF.

Today, the list of targets and tools has increased. The two major options are the targeting of the ligand or the receptor [42] (Fig. 4). Regarding the ligand, the possible choice is between IL-17A or IL-17F or both, and for the receptor, IL-17RA or IL-17RC. In addition, administration of IL-25, the member of the IL-17 family with opposite effects, may be of interest because of its Th2 properties to inhibit, at the same time, IL-17, IL-1, TNF, and IL-6. These results show similarities with the effects of IL-4. In addition, as for the other proinflammatory cytokines, active research is looking for small molecules able to control the intracellular signaling pathways.

Targeting cytokines may interfere with immune defense. Blocking TNF was associated with an increased risk of tuberculosis reactivation. The mechanism implies an effect on the Th1 pathway and on the induction of IFN-γ by IL-12 and IL-18 [18]. For IL-17, a link with neutrophils was apparent from the first results. This implies that inhibition of IL-17 may have consequences on the acute defense mechanisms involving neutrophils. Indeed, in the mouse, inhibition of the IL-17 system has been associated with increased mortality from bacterial lung infections [43]. IL-17 appears to be critical for neutrophil activation and migration [44]. IL-17 is a strong inducer of IL-8, a key chemokine for neutrophils. Conversely, IL-17 appears to have inhibitory effect on the production of other chemokines involved in the migration of mononuclear cells. In addition, inhibition of the IL-23 pathway has been associated with defects in the cell-mediated immunity including increased severity of mycobacterial infections.

The position of IL-17 inhibition in the treatment of RA and other inflammatory conditions remains to be defined. Coming back to the synergistic interactions, an enhanced inhibitory activity was observed with the combination of TNF and IL-17 inhibitors using *ex vivo* samples of RA synovium and bone [45]. Thus, primary or secondary lack of response to TNF inhibitors may represent a useful addition. It is possible that the combined inhibition of TNF and IL-17 may have the advantage of targeting two different cell types, monocytes and T cells. This would also control

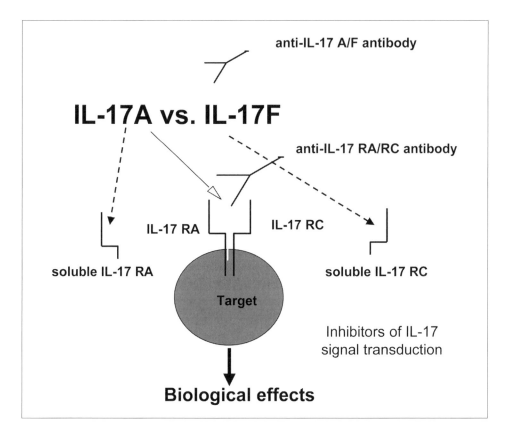

Figure 4
Modalities to inhibit IL-17 action.

the enhancing effects of IL-17 on TNF production by monocytes. However, only properly designed clinical trials could address this issue.

Anti-TNF non-responders may have an IL-17-driven disease or secondary loss of response to TNF inhibition may result from the induction or use of other pathways, possibly involving IL-17, taking over the initial predominant TNF contribution. Results in the mouse are in line with this possibility [11].

Conclusion

The story of IL-17 started 10 years ago and this is the time it took to become a cytokine in fashion. The identification of the Th17 subset indicates that some T cells are involved in and amplify the link between chronic inflammation and extracellular matrix destruction. Similar concepts apply to other complex diseases with inflammation-induced destruction, such as multiple sclerosis and Crohn's disease where the contribution of IL-17 has already been identified. Tools are now almost ready to verify whether these concepts, already 10 years old, are indeed correct.

Acknowledgements

I would like to thank all contributors to the IL-17 studies from our group over the years: Martine Chabaud, Masanori Kawashima, Corinne Granet, Guillaume Chevrel, Guillaume Page, Yuan Zhou, and today, Ling Toh and Saloua Zrioual.

References

1   Miossec P (2003) Interleukin-17 in rheumatoid arthritis: If T cells were to contribute to inflammation and destruction through synergy. *Arthritis Rheum* 48: 594–601

2   Fossiez F, Djossou O, Chomarat P, Flores-Romo L, Ait-Yahia S, Maat C, Pin JJ, Garrone P, Garcia E, Saeland S et al (1996) T cell interleukin-17 induces stromal cells to produce proinflammatory and hematopoietic cytokines. *J Exp Med* 183: 2593–603

3   Yao Z, Painter SL, Fanslow WC, Ulrich D, Macduff BM, Spriggs MK, Armitage RJ (1995) Human IL-17: A novel cytokine derived from T cells. *J Immunol* 155: 5483–6

4   Koenders MI, Joosten LA, van den Berg WB (2006) Potential new targets in arthritis therapy: Interleukin (IL)-17 and its relation to tumour necrosis factor and IL-1 in experimental arthritis. *Ann Rheum Dis* 65 (Suppl 3): iii29–iii33

5   Chabaud M, Lubberts E, Joosten L, van Den Berg W, Miossec P (2001) IL-17 derived from juxta-articular bone and synovium contributes to joint degradation in rheumatoid arthritis. *Arthritis Res* 3: 168–77

6   Lubberts E, Joosten LA, van de Loo FA, Schwarzenberger P, Kolls J, van den Berg WB (2002) Overexpression of IL-17 in the knee joint of collagen type II immunized mice promotes collagen arthritis and aggravates joint destruction. *Inflamm Res* 51: 102–4

7   Lubberts E, Joosten LA, Chabaud M, van Den Bersselaar L, Oppers B, Coenen-De Roo CJ, Richards CD, Miossec P, van Den Berg WB (2000) IL-4 gene therapy for collagen arthritis suppresses synovial IL-17 and osteoprotegerin ligand and prevents bone erosion. *J Clin Invest* 105: 1697–710

8   Chabaud M, Durand JM, Buchs N, Fossiez F, Page G, Frappart L, Miossec P (1999) Human interleukin-17: A T cell-derived proinflammatory cytokine produced by the rheumatoid synovium. *Arthritis Rheum* 42: 963–70

9   Sato K, Suematsu A, Okamoto K, Yamaguchi A, Morishita Y, Kadono Y, Tanaka S, Kodama T, Akira S, Iwakura Y et al (2006) Th17 functions as an osteoclastogenic helper T cell subset that links T cell activation and bone destruction. *J Exp Med* 203: 2673–82

10 Page G, Miossec P (2005) RANK and RANKL expression as markers of dendritic cell-T cell interactions in paired samples of rheumatoid synovium and lymph nodes. *Arthritis Rheum* 52: 2307–12

11 Koenders MI, Lubberts E, van de Loo FA, Oppers-Walgreen B, van den Bersselaar L, Helsen MM, Kolls JK, Di Padova FE, Joosten LA, van den Berg WB (2006) Interleukin-17 acts independently of TNF-alpha under arthritic conditions. *J Immunol* 176: 6262–9

12 Chabaud M, Fossiez F, Taupin JL, Miossec P (1998) Enhancing effect of IL-17 on IL-1–induced IL-6 and leukemia inhibitory factor production by rheumatoid arthritis synoviocytes and its regulation by Th2 cytokines. *J Immunol* 161: 409–14

13 Jovanovic DV, Di Battista JA, Martel-Pelletier J, Jolicoeur FC, He Y, Zhang M, Mineau F, Pelletier JP (1998) IL-17 stimulates the production and expression of proinflammatory cytokines, IL-beta and TNF-alpha, by human macrophages. *J Immunol* 160: 3513–21

14 Harrington, LE, Mangan PR, Weaver CT (2006) Expanding the effector CD4 T-cell repertoire: The Th17 lineage. *Curr Opin Immunol* 18: 349–56

15 Iwakura Y, Ishigame H (2006) The IL-23/IL-17 axis in inflammation. *J Clin Invest* 116: 1218–22

16 Weaver CT, Harrington LE, Mangan PR, Gavrieli M, Murphy KM (2006) Th17: An effector CD4 T cell lineage with regulatory T cell ties. *Immunity* 24: 677–88

17 Bettelli E, Korn T, Oukka M, Kuchroo VK (2008) Induction and effector functions of T(H)17 cells. *Nature* 453: 1051–7

18 Kawashima M, Miossec P (2004) Decreased response to IL-12 and IL-18 of peripheral blood cells in rheumatoid arthritis. *Arthritis Res Ther* 6: R39–R45

19 Cua DJ, Sherlock J, Chen Y, Murphy CA, Joyce B, Seymour B, Lucian L, To W, Kwan S, Churakova T et al (2003) Interleukin-23 rather than interleukin-12 is the critical cytokine for autoimmune inflammation of the brain. *Nature* 421: 744–8

20 Trinchieri G (2003) Interleukin-12 and the regulation of innate resistance and adaptive immunity. *Nat Rev Immunol* 3: 133–46

21 Chen Y, Langrish CL, McKenzie B, Joyce-Shaikh B, Stumhofer JS, McClanahan T, Blumenschein W, Churakovsa T, Low J, Presta L et al (2006) Anti-IL-23 therapy inhibits multiple inflammatory pathways and ameliorates autoimmune encephalomyelitis. *J Clin Invest* 116: 1317–1326

22 Ivanov II, McKenzie BS, Zhou L, Tadokoro CE, Lepelley A, Lafaille JJ, Cua DJ, Littman DR (2006) The orphan nuclear receptor RORgammat directs the differentiation program of proinflammatory IL-17+ T helper cells. *Cell* 126: 1121–33

23 Rangachari M, Mauermann N, Marty RR, Dirnhofer S, Kurrer MO, Komnenovic V, Penninger JM, Eriksson U (2006) T-bet negatively regulates autoimmune myocarditis by suppressing local production of interleukin 17. *J Exp Med* 203: 2009–19

24 Sutton C, Brereton C, Keogh B, Mills KH, Lavelle EC (2006) A crucial role for interleukin (IL)-1 in the induction of IL-17-producing T cells that mediate autoimmune encephalomyelitis. *J Exp Med* 203: 1685–91

25 Luger D, Silver PB, Tang J, Cua D, Chen Z, Iwakura Y, Bowman EP, Sgambellone NM, Chan CC, Caspi RR (2008) Either a Th17 or a Th1 effector response can drive autoimmunity: Conditions of disease induction affect dominant effector category. *J Exp Med* 205: 799–810

26 Kroenke MA, Carlson TJ, Andjelkovic AV, Segal BM (2008) IL-12- and IL-23-modulated T cells induce distinct types of EAE based on histology, CNS chemokine profile, and response to cytokine inhibition. *J Exp Med* 205: 1535–41

27 Aarvak T, Chabaud M, Miossec P, Natvig JB (1999) IL-17 is produced by some proinflammatory Th1/Th0 cells but not by Th2 cells. *J Immunol* 162: 1246–51

28 Page G, Sattler A, Kersten S, Thiel A, Radbruch A, Miossec P (2004) Plasma cell-like morphology of Th1-cytokine-producing cells associated with the loss of CD3 expression. *Am J Pathol* 164: 409–17

29 Mangan PR, Harrington LE, O'Quinn DB, Helms WS, Bullard DC, Elson CO, Hatton RD, Wahl SM, Schoeb TR, Weaver CT (2006) Transforming growth factor-beta induces development of the T(H)17 lineage. *Nature* 441: 231–4

30 Bettelli E, Carrier Y, Gao W, Korn T, Strom TB, Oukka M, Weiner HL, Kuchroo VK (2006) Reciprocal developmental pathways for the generation of pathogenic effector TH17 and regulatory T cells. *Nature* 441: 235–8

31 Sarkar S, Tesmer LA, Hindnavis V, Endres JL, Fox DA (2006) Interleukin-17 as a molecular target in immune-mediated arthritis: Immunoregulatory properties of genetically modified murine dendritic cells that secrete interleukin-4. *Arthritis Rheum* 56: 89–100

32 Annunziato F, Cosmi L, Santarlasci V, Maggi L, Liotta F, Mazzinghi B, Parente E, Fili L, Ferri S, Frosali F et al (2007) Phenotypic and functional features of human Th17 cells. *J Exp Med* 204: 1849–61

33 Yang L,,erson DE, Baecher-Allan C, Hastings WD, Bettelli E, Oukka M, Kuchroo VK, Hafler DA (2008) IL-21 and TGF-beta are required for differentiation of human T(H)17 cells. *Nature* 454: 350–2

34 Zrioual S, Toh ML, Tournadre A, Zhou Y, Cazalis MA, Pachot A, Miossec V, Miossec P (2008) IL-17RA and IL-17RC receptors are essential for IL-17A-induced ELR$^+$ CXC chemokine expression in synoviocytes and are overexpressed in rheumatoid blood. *J Immunol* 180: 655–63

35 Starnes T, Broxmeyer HE, Robertson MJ, Hromas R (2002) Cutting edge: IL-17D, a novel member of the IL-17 family, stimulates cytokine production and inhibits hemopoiesis. *J Immunol* 169: 642–6

36 Owyang AM, Zaph C, Wilson EH, Guild KJ, McClanahan T, Miller HR, Cua DJ, Goldschmidt M, Hunter CA, Kastelein RA, Artis D (2006) Interleukin 25 regulates type 2 cytokine-dependent immunity and limits chronic inflammation in the gastrointestinal tract. *J Exp Med* 203: 843–9

37 Liang SC, Tan XY, Luxenberg DP, Karim R, Dunussi-Joannopoulos K, Collins M, Fouser LA (2006) Interleukin (IL)-22 and IL-17 are coexpressed by Th17 cells and cooperatively enhance expression of antimicrobial peptides. *J Exp Med* 203: 2271–9

38 Zheng Y, Danilenko DM, Valdez P, Kasman I, Eastham-Anderson J, Wu J, Ouyang W

(2007) Interleukin-22, a T(H)17 cytokine, mediates IL-23-induced dermal inflammation and acanthosis. *Nature* 445: 648–51

39   Yao Z, Fanslow WC, Seldin MF, Rousseau AM, Painter SL, Comeau MR, Cohen JI, Spriggs MK (1995) Herpesvirus Saimiri encodes a new cytokine, IL-17, which binds to a novel cytokine receptor. *Immunity* 3: 811–21

40   Yao Z, Spriggs MK, Derry JM, Strockbine L, Park LS, VandenBos T, Zappone JD, Painter SL, Armitage RJ (1997) Molecular characterization of the human interleukin (IL)-17 receptor. *Cytokine* 9: 794–800

41   Toy D, Kugler D, Wolfson M, Bos TV, Gurgel J, Derry J, Tocker J, Peschon J (2006) Cutting edge: Interleukin 17 signals through a heteromeric receptor complex. *J Immunol* 177: 36–9

42   Dong C (2008) IL-23/IL-17 biology and therapeutic considerations. *J Immunotoxicol* 5: 43–6

43   Dubin PJ, Kolls JK (2007) IL-23 mediates inflammatory responses to mucoid Pseudomonas aeruginosa lung infection in mice. *Am J Physiol Lung Cell Mol Physiol* 292: L519–28

44   Chung DR, Kasper DL, Panzo RJ, Chtinis T, Grusby MJ, Sayegh MH, Tzianabos AO (2003) CD4$^+$ T cells mediate abscess formation in intra-abdominal sepsis by an IL-17-dependent mechanism. *J Immunol* 170: 1958–63

45   Chabaud M, Miossec P (2001) The combination of tumor necrosis factor alpha blockade with interleukin-1 and interleukin-17 blockade is more effective for controlling synovial inflammation and bone resorption in an *ex vivo* model. *Arthritis Rheum* 44: 1293–303

Role of IL-18 in inflammatory diseases

Charles A. Dinarello

Department of Medicine, Division of Infectious Diseases, University of Colorado Denver, 12700 East 19th Ave, B168, Aurora, CO 80045, USA

Abstract

IL-18 is a member of the interleukin (IL)-1 family. IL-1β and IL-18 are closely related, and both require the intracellular cysteine protease caspase-1 for biological activity. Several autoimmune diseases are thought to be mediated, in part, by IL-18. Many are those with associated elevated interferon-γ (IFN-γ) levels, such as systemic lupus erythematosus, macrophage activation syndrome, rheumatoid arthritis, Crohn's disease, psoriasis and graft *versus* host disease. In addition, ischemia, including acute renal failure in humans, appears to involve IL-18. Animal studies also support the concept that IL-18 is a key player in models of lupus erythematosus, atherosclerosis, graft versus host disease and hepatitis. Unexpectedly, IL-18 plays a role in appetite control and the development of obesity. The IL-18-binding protein, a naturally occurring, specific inhibitor of IL-18, neutralizes IL-18 activities and has been shown to be safe in patients. Other options for reducing IL-18 activities are inhibitors of capsase-1, human monoclonal antibodies to IL-18, soluble IL-18 receptors and anti-IL-18 receptor monoclonal antibodies.

Introduction

IL-18 is a member of the IL-1 family of cytokines and is structurally related to IL-1β [1]. Recently, a new member of the IL-1 family, IL-33, has been reported; structurally IL-33 is closely related to IL-18 [2]. However, unlike IL-18, IL-33 binds to its own receptor, ST2, a long-time orphan receptor in the IL-1 family of cytokines [2]. The IL-1β and the IL-18 precursors require caspase-1 for cleavage, activity and release [3–5]. Therefore, anti-proteases that inhibit capase-1 reduce both the processing and release of IL-1β and IL-18. Now, IL-33 can be added to the list of members of the IL-1 family that require caspase-1 for processing and release [2]. However, it is important to note that IL-18 is not a recapitulation of the biology or clinical significance of IL-1, or its activity similar to the biologically activity of IL-33; in fact, IL-18 is a unique cytokine exhibiting inflammatory as well as immunoregulatory processes distinct from IL-1β or IL-33. For example, IL-1β is not required for IFN-γ production, whereas IL-18 is [6]. Initially thought of as primarily a Th1-polarizing cytokine, IL-18 is also relevant to Th2 diseases [7]. As discussed in this review, animal models reveal that targeting IL-18 holds promise for the treatment of autoimmune and inflammatory diseases.

IL-18 as an immunoregulatory cytokine

The importance of IL-18 as an immunoregulatory cytokine is derived from its prominent biological property of inducing IFN-γ. IL-18 was first described in 1989 in the serum following an injection of endotoxin into mice pretreated with *Proprionibacterium acnes* and shown to induce IFN-γ; however, at that time many investigators concluded that the serum factor was nothing but IL-12. With a great deal of diligence, the putative IFN-γ-inducing factor activity was purified from thousands of mouse livers and the N-terminal amino acid sequence revealed a unique cytokine, not IL-12. With molecular cloning of this "IFN-γ-inducing factor" in 1995 [1], the name was changed to IL-18. Surprisingly, the new cytokine was related to IL-1 and particularly to IL-1β. Both cytokines, lacking signal peptides, are first synthesized as inactive precursors, and neither is secreted *via* the Golgi. Following cleavage by caspase-1, the active "mature" cytokines are released. Macrophages and dendritic cells are the primary sources for active IL-18, but the IL-18 precursor is found constitutively expressed in epithelial cells throughout the body. Previously, it was thought that inhibition of caspase-1 as a therapeutic target was specific for reducing the activity of IL-1β, but it became clear that IL-18 activity would also be affected. In fact, any phenotypic characteristic of capsase-1-deficient mice undergoing inflammatory challenges must be differentiated as due to reduced IL-1β or IL-18 activity. For example, the IL-1β-deficient mouse is susceptible to models of colitis, whereas the caspase-1-deficient mouse is resistant [8]; antibodies to IL-18 are protective, whereas the IL-1 receptor antagonist is not [8, 9].

Because of its role in the production of IFN-γ, T cell polarization is a characteristic of IL-18 whereas IFN-γ induction is not a prominent characteristic of IL-1. IL-18 exhibits characteristics of other pro-inflammatory cytokines, such as increases in cell adhesion molecules, nitric oxide synthesis, and chemokine production. A unique property of IL-18 is the induction of Fas ligand (FasL). The induction of fever, an important clinical property of IL-1, TNF-α and IL-6, is not a property of IL-18. Injection of IL-18 into mice, rabbits or in humans does not produce fever [10, 11]. Unlike IL-1 and TNF-α, IL-18 does not induce cyclooxygenase-2 and hence there is no production of prostaglandin E2 [12, 13]. IL-18 has been administered to humans for the treatment of cancer to increase the activity and expansion of cytotoxic T cells. Although the results of clinical trials are presently unknown, several preclinical studies reveal the benefit of IL-18 administration in certain models of rodent cancer. Not unexpectedly and similar to several cytokines, the therapeutic focus on IL-18 has shifted from its use as an immune stimulant to inhibition of its activity.

Because IL-18 can increase IFN-γ production, blocking IL-18 activity in autoimmune diseases is potentially an attractive therapeutic target. However, anti-IL-12 has been shown to reduce the severity of Crohn's Disease as well as psoriasis. Therefore, IL-12 can induce IFN-γ in the absence of IL-18. However, there are many models of IL-18 activity independent of IFN-γ. For example, we have recently reported a new cytokine, IL-32, which was discovered in the total absence of IL-12 or IFN-γ [14]. Furthermore, models of inhibition of proteoglycan synthesis are IL-18 dependent but IFN-γ independent [15]. In addition, IL-18-dependent melanoma metastasis to the liver is IFN-γ independent [16]. The results of preclinical studies and the targeting of IL-18 to treat autoimmune and inflammatory diseases are discussed in this article.

Therapeutic strategies for reducing IL-18 activities

The strategies for reducing IL-18 activity include neutralizing anti-IL-18 monoclonal antibodies, caspase-1 inhibitors and blocking antibodies to the IL-18 receptor (IL-18R) chains. Caspase-1 inhibitors are oral agents and are presently in clinical trials in rheumatoid arthritis; a reduction in the signs and symptoms of the disease has been observed. Caspase-1 inhibitors prevent the release of active IL-1β and IL-18 and, therefore, may derived clinical benefit by reducing the activities of both cytokines [3, 4, 17]. A naturally occurring IL-18-binding protein (IL-18BP) was discovered in 1999; IL-18BP is effective in neutralizing IL-18 activity [18]. IL-18BP is not a soluble form of either chain of the IL-18R but rather a constitutively secreted, high-affinity and specific inhibitor of IL-18 [19, 20]. IL-18BP is currently in clinical trials for the treatment of rheumatoid arthritis and severe psoriasis. The pharmacokinetics of IL-18BP have been reported and IL-18BP is safe even at the highest doses in over 6 weeks of treatment [21].

Caspase-1 and non-caspase-1 processing of IL-18

The importance of caspase-1 in inflammation has been revealed in patients with mutations in the NALP3 gene locus, which participates in the conversion of procaspase-1 to active caspase-1. Single amino acid point mutations in the gene product result in increased processing and release of IL-1β [22]. Clinical manifestations include mental retardation, hearing loss, exquisite sensitivity to cold and deforming arthritis [23]. In some patients who had extremely high levels of serum amyloid A protein with renal deposits and terminal renal failure, a near total reversal in both the symptoms and biochemical abnormalities of the disease were seen within a few days of IL-1 blockade using the IL-1 receptor antagonist [24]. It is likely that IL-18 also contributes to disease in these patients.

The non-caspase-1 enzyme associated with processing both the IL-1 and IL-18 precursors is proteinase-3 (PR-3) [25]. Agonistic autoantibodies to PR-3 are pathological in Wegener's granulomatosis and may contribute to the non-caspase-1 cleavage of the IL-18 precursor and IFN-γ production in this disease. Epithelial cells stimulated with PR-3 in the presence of endotoxin release active IL-18 into the supernatant [26]. Since lactate dehydrogenase activity is not released, the appearance of active IL-18 is not due to cell leakage or death. Injecting mice with recombinant FasL results in hepatic damage that is IL-18 dependent [27]. However, FasL-mediated cell death is IL-18 dependent and caspase-1 independent [27], but ischemia-reperfusion injury results is cell death is *via* an IL-18- as well as a caspase-1-dependent pathway [28, 29].

P2X7 receptor targeting

The P2X7 receptor is involved in the secretion of IL-1β as well as IL-18 [30–32]. Stimulation of this receptor by ATP is a well-described event in the release of IL-1β and IL-18. A tyrosine derivative named KN-62 exhibits selective P2X7 receptor-blocking properties [33]. In a study of small molecule inhibitors of this receptor, analogues of KN-62-related compounds were characterized for their ability to affect the human P2X7 receptor on monocyte-derived human macrophages [33]. Although several analogues inhibited the secretion of IL-1β, no data exist on the effect of these inhibitors on IL-18 secretion [33]. Unlike IL-1β, the secretion of IL-18 is mostly studied *in vivo* in mice that have been treated with *Cryptosporidium parvum* [4], rather than *in vitro*. *In vitro*, the release of IL-18 requires the presence of activated T cells [34, 35].

Targeting the IL-18Rs

Antibodies to either chain of the IL-18R complex are attractive options for treating IL-18-mediated diseases. The IL-18R chains (IL-18Rα and IL-18Rβ) are members of the IL-1 receptor family. The binding sites for IL-18 to the IL-18Rα chain are similar to those for IL-1 binding to the IL-1 receptor type I [36–38]. Two sites bind to the ligand binding chain (IL-18Rα) and a third site binds to the IL-18Rβ chain, also called the signal transducing chain. The intracellular chains of the IL-18Rs contain the Toll domains, which are essential for initiating signal transduction (see Fig. 1). The Toll domains of the IL-18Rs are similar to the same domains of the Toll-like receptors, which recognize various microbial products, viruses and nucleic acids. As a therapeutic option, however, commercial antibodies generated to the IL-18Rα and β chains are 100-fold less effective in neutralizing IL-18 activity compared to the IL-18BP [39]. Nevertheless, the development of blocking antibodies to IL-18R

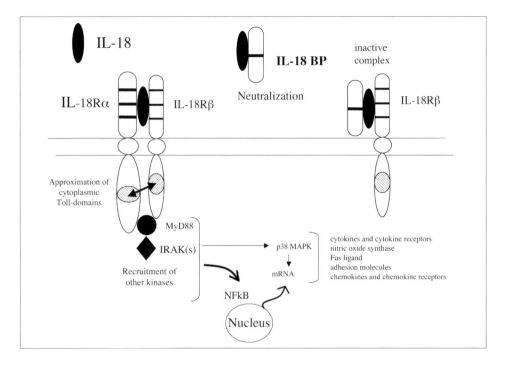

Figure 1

IL-18 activation of cell signaling. Mature IL-18 binds to the IL-18Rα chain and recruits the IL-18Rβ chain, resulting in the formation of a heterodimeric complex. As a result of the formation of the extracellular complex, the intracellular chains also form a complex, which brings the Toll domains of each receptor chain into close proximity. Although poorly under-stood, the close proximity of the Toll domains recruits the intracellular protein MyD88 to the receptor chains. MyD88 is common to cells activated by IL-1, IL-18 and TLR-4 ligands (endotoxins). Following MyD88 recruitment, there is a rapid phosphorylation of the IL-1 receptor-activating kinases (IRAK). There are four IRAK proteins. Depending on the cell type, other kinases have been reported to undergo phosphorylation. These are the TNF recep-tor activating factor (TRAF)-6 and inhibitory kappa B kinases (IKK) α and β (not shown). Phosphorylation of IKK results in the phosphorylation of IκB and translocation of NF-κB to the nucleus. However, this is not observed uniformly in all cell types and there are distinct differences in NF-κB activation in different cells stimulated with IL-18 [13]. In addition, IL-18-activated cells phosphorylate mitogen-activated protein kinase (MAPK) p38. In IL-18-activated cells, new genes are expressed and translated. Those shown in the figure represent the pro-inflammatory genes. The presence of IL-18BP prevents IL-18-induced cellular activa-tion. IL-18BP is present in the extracellular milieu as a constitutively expressed protein where it can bind and neutralize IL-18, thus preventing activation of the cell surface receptors. In addition, formation of inactive complexes of IL-18BP with IL-18 and the IL-18Rβ chain deprives the cell of the participation of IL-18Rβ chain in activating the cell.

chains remains a viable therapeutic option since an antibody to the type I IL-1 receptor chain is in clinical trials in rheumatoid arthritis.

Unless converted into a fusion protein in somewhat the same manner as that for other soluble cytokine receptors, it is unlikely that the soluble form of the monomeric form of the IL-18Rα is a candidate therapeutic agent due to its low affinity. Another member of the IL-1 family (IL-1F), IL-1F7 [40], may be the naturally occurring receptor antagonist of IL-18. IL-1F7 binds to the IL-18Rα chain with a high affinity but this binding does not recruit the IL-18Rβ chain. The occupancy of the IL-18Rα without formation of the heterodimer with the IL-18Rβ is the same mechanism by which the IL-1 receptor antagonist prevents the activity of IL-1. However, IL-1F7 does not affect the activity of IL-18 [41, 42] and the biological significance of IL-1F7 binding to the IL-18Rα remains unclear. However, in the presence of low concentrations of IL-18BP, IL-1F7 has been shown to reduce the activity of IL-18 [43].

IL-18BP

The discovery of the IL-18BP occurred during the search for the extracellular (soluble) receptors for IL-18 in human urine. Nearly all the soluble cytokine receptors are found in human urine [44]. For example, the TNF p75 soluble receptor, used widely for the treatment of rheumatoid arthritis, ankylosing spondylitis and psoriasis, was initially purified and sequenced using ligand-specific affinity chromatography [45]. In searching for soluble IL-18 receptors, IL-18 was covalently bound to a matrix and highly concentrated human urine, donated by Italian nuns, was passed over the matrix and eluted with acid to disrupt the ligand (in this case IL-18) for its soluble receptors. Unexpectedly, instead of the elution of soluble forms of the cell surface IL-18Rs, the IL-18BP was discovered [18]. This was due to the higher affinity of the IL-18BP for the ligand compared to the soluble receptors.

The IL-18BP is a constitutively secreted protein, with a high affinity (400 pM) binding to IL-18. There is very limited amino acid sequence homology between IL-18BP and the cell surface IL-18Rs; IL-18BP lacks a transmembrane domain and contains only one Ig-like domain [20, 46]. IL-18BP shares many characteristics with the soluble form of the IL-1 type II receptor in that both function as decoys to prevent the binding of their respective ligands to the signaling receptor chains [47]. The fact that there is limited amino acid homology between IL-18BP and the IL-1 receptor type II suggests a common ancestor. In humans, IL-18BP is highly expressed in spleen and the intestinal tract, both immunologically active tissues [18]. Alternate mRNA splicing of IL-18BP results in four isoforms [18, 20]. Of considerable importance is that the prominent 'a' isoform is present in the serum of healthy humans at a 20-fold molar excess compared to IL-18 [19]. This level of

IL-18BP may contribute to a default mechanism by which a Th1 response to foreign organisms is blunted to reduce triggering an autoimmune responses to a routine infection. The promoter for IL-18BP contains two IFN-γ response elements [48] and constitutive gene expression for IL-18BP is IFN-γ dependent [49], suggesting a compensatory feedback mechanism. Thus, elevated levels of IFN-γ stimulate more IL-18BP in an attempt to reduce IL-18-mediated IFN-γ production. For example, in mice deficient in IFN regulatory factor-1, a transcription factor for IFN-γ, low to absent tissue levels of IL-18BP are found compared to wild-type mice [50]. These IFN regulatory factor-1-deficient mice are exquisitely sensitive to colitis, but when treated with exogenous IL-18BP exhibit reduced disease [51].

Viral IL-18BP

The most convincing evidence that IL-18 is a major player in inflammatory conditions and that IL-18BP is functional in combating inflammation comes from a natural experiment in humans. *Molluscum contagiosum* is a common viral infection of the skin often seen in children and individuals with HIV-1 infection. The infection is characterized by raised but bland eruptions; there are large numbers of viral particles in the epithelial cells of the skin but histologically there are few inflammatory or immunologically active cells in or near the lesions. Clearly, the virus fails to elicit an inflammatory or immunological response. A close amino acid similarity exists between human IL-18BP and a gene found in various members of the Poxviruses. The greatest homology is with in *M. contagiosum* [18, 52, 53]. The viral genes encoding for viral IL-18BP have been expressed and the recombinant proteins neutralize mammalian IL-18 activity [52, 53]. The ability of viral IL-18BP to reduce the activity of mammalian IL-18 likely explains the lack of inflammatory and immune cells in the infected skin and the blandness of the lesions. One can conclude from this natural experiment of *M. contagiosum* infection that blocking IL-18 reduces immune and inflammatory processes such as the function of dendritic and inflammatory cells.

IL-18:IL-18BP imbalance in macrophage-activating syndrome

The macrophage-activating syndrome (MAS), also known as hemophagocytic syndrome, is characterized by an uncontrolled and poorly understood activation of Th1 lymphocytes and macrophages. In a study of 20 patients with MAS secondary to infections, autoimmune disease, lymphoma, or cancer, the concentrations of circulating IL-18, IL-18BP, IFN-γ and IL-12 were determined and matched with clinical parameters. In the MAS, but not in control, patients there was evidence of highly increased stimulation of macrophages and natural killer (NK) cells. Most importantly, concentrations of IL-18BP were only moderately elevated, resulting in a high level of biologically active free IL-18 [19] in MAS (4.6-fold increase compared with controls ($p < 0.001$). Others have reported marked expression of IL-18 in fatal MAS [54]. The concentrations of free IL-18 but not IL-12 significantly correlated with clinical status and the biological markers of MAS such as anemia, hypertriglyceridemia and hyperferritinemia, and also with markers of Th1 lymphocyte or macrophage activation such as elevated concentrations of IFN-γ, soluble IL-2 and TNF receptors. Therefore, treatment of life-threatening MAS with IL-18BP is a logical therapeutic intervention to correct the severe IL-18:IL-18BP imbalance resulting in Th1 lymphocyte and macrophage activation.

Neutralizing antibodies to IL-18

Although there are no clinical trials of neutralizing antibodies to IL-18, pre-clinical studies have employed IL-18 antibodies to reduce IL-18 activity in animal models of disease. The results of these studies are shown in Table 1. Assuming that neutralizing antibodies to IL-18 are developed and tested in human diseases, what are the anticipated differences between a neutralizing antibody and a neutralizing soluble receptor or a binding protein such as IL-18BP? First, to evaluate such differences, the agent with the highest affinity is preferable. From a pharmacokinetic viewpoint, a long half-life is preferable. One can increase the binding affinity for a ligand by converting a soluble receptor or binding protein to a divalent fusion protein. However, the danger here is the increase risk of creating a novel epitope for antibody production. The advantage of monoclonal antibodies is that they are human and the risk of developing antibodies to a human antibody is reduced significantly. At first glance, one would conclude that high affinity, human antibodies to IL-18 are preferable to the IL-18BP. However, if a divalent fusion protein of IL-18BP has a high affinity and is not immunogenic, the next issue is a comparison of the half-life of a monoclonal antibody to that of a fusion protein. Here the issue is one of safety. A short half-life is preferential for rapid cessation of therapy in the event of a life-threatening infection, whereas long half-life antibodies exert their effects of suppressing host defense for weeks. In fact, the large body of evidence for comparing infections associated with anti-TNF-α monoclonal antibodies (infliximab or adalimumab) to the soluble TNF p75 receptor fusion protein (etanercept) shows that the differences may, in part, be due to differences in half-life as well as mechanism of action [55]. In the case of neutralizing IL-18, the suppression of IFN-γ is of concern for host defense against intracellular organisms such as *Mycobacterium tuberculosis* [56].

Table 1 - Reduction in disease severity with blocking of endogenous IL-18 activity.

Disease model	Intervention	Outcome
Acute DSS-induced colitis	Anti-IL-18 antibodies [9]; IL-18BP [90]	↓ clinical disease; ↓ TNF-α, IFN-γ, IL-1, MIP-1,2
Chronic DSS-induced colitis	Caspase-1 KO [8]	↓ IL-1β, IFN-γ and CD3 cells
TNBS colitis colitis-induced colitis	IL-18BP [91]	↓ clinical disease; ↓ cytokines
CD62/CD4 T-cell-induced colitis	Adenoviral antisense IL-18 [92]	↓ clinical disease; ↓ mucosal IFN-γ
Streptococcal wall-induced arthritis	IL-18 antibodies [15]	↑ cartilage proteoglycan synthesis, ↓ inflammation;
Collagen-induced arthritis	IL-18BP; IL-18 [57]; Ad-viral IL-18BP [93]	↓ clinical disease; ↓ cytokines
Collagen-induced arthritis	IL-18 deficient mice [56]	↓ clinical disease; ↓ cytokines
Graft *versus* host disease	Anti-18 [63]	↓ CD8⁺-mediated mortality
Lupus prone mice	IL-18 vaccination [58]	↓ mortality; ↓ nephritis
Allergic airway hyperresponsiveness	IL-18 vaccination [94]	↓ bronchocontriction
Experimental myasthenia gravis	Anti- IL-18 [95]	↓ clinical disease
Autoimmune encephalomyelitis	Caspase-1 KO [96]; caspase-1 inhibition [96]	↓ clinical disease; ↓ IFN-γ
Con-A-induced hepatitis	Anti-18 [97]; IL-18BP [26]; IL-18BP-Tg [98]	↓ liver enzymes
Fas-mediated hepatic failure	IL-18 deficient mice [99]; IL-18BP [26]	↓ liver necrosis
Pseudomonas exotoxin-A hepatitis	IL-18BP [26]	↓ liver enzymes; ↓ IFN-γ
IL-12-induced IFN-γ	Anti-18 [100]; caspase-1 KO [100]	↓ IFN-γ
Endotoxin-induced IFN-γ	Anti-18 [1, 101]; IL-18BP [18, 26]; caspase-1 KO	↓ IFN-γ
LPS-induced hepatic necrosis	Anti-IL-18 monoclonal [1, 101]; IL-18BP [26]	↓ necrosis; ↓ TNF-α; ↓ FasL
LPS-induced lung neutrophils	IL-18 [101]	↑ survival; ↓ myeloperoxidase

Table 1 (continued)

Disease model	Intervention	Outcome
Melanoma hepatic metastasis	IL-18BP [16, 102]	↓ metastatic foci; ↓ VCAM-1
Ischemia-induced hepatic failure	Anti-18 [103]	↓ apoptosis; ↓ NF-κB
Ischemia-induced acute renal failure	Anti-IL-18 polyclonal [28]; caspase-1 KO [28]	↓ creatinine; ↓ urea
Ischemic myocardial dysfunction	IL-18BP [27]; caspase-1 inhibition [27]	↑ myocardial contractility
LPS-induced myocardial suppression	Anti-IL-18 polyclonal [77]	↑ heart contractility; ↓ IL-1β
Atherosclerosis in ApoE KO mice	IL-18BP [82]	↓ plaques; ↓ infiltrates, ↑ vessel collagen

DSS, dextran sulphate sodium; ApoE, apolipoprotein E; KO, knockout

Blocking IL-18 in disease models

As with any cytokine, the role of IL-18 in a particular disease process is best assessed employing specific neutralization of the cytokine in a complex disease model. Although mice deficient in IL-18 have been generated and tested for the development of autoimmune diseases [57], any reduction in severity may be due to a reduction in the immune response such as to antigens or the sensitization processes itself and does not address the effect of IL-18 on established disease. IL-18 neutralization in wild-type mice is effective in reducing collagen-induced arthritis [58] as well as inflammatory arthritis [15]. Inflammatory arthritis is of particular relevance since this is a model of cartilage loss due to decreased proteoglycan synthesis and is independent of IFN-γ. IL-18 contributes to the lupus-like disease in mice [59] *via* IFN-γ production. Caspase-1-deficient mice provide useful models for disease [6, 29] but here the effect may be on IL-1β, IL-18 or both.

Most investigations initially focused on IL-18 in Th1-mediated diseases in which IFN-γ plays a prominent role. However, it soon became clear that blocking IL-18 resulted in reduction of disease severity in models where IFN-γ has no significant role or in mice deficient in IFN-γ. For example, IL-18-mediated loss of cartilage synthesis in arthritis models is IFN-γ independent [15]. Prevention of melanoma metastases is IL-18 dependent but IFN-γ independent [16], and similar findings exist for ischemia-reperfusion injury in the heart, kidney and liver. Table 1 lists various animal models of Th1-, Th2- as well as non-immune-mediated disease where the effect of reducing endogenous IL-18 activities has been reported.

IL-18 in Th1-like diseases

In driving the Th1 response, IL-18 appears to act in association with IL-12 or IL-15, as IL-18 alone does not induce IFN-γ. The effect of IL-12 is, in part, to increase the expression of IL-18Rs on T lymphocytes, thymocytes and NK cells [7, 60, 61]. It appears that the role of IL-18 in the polarization of the Th1 response is dependent on IFN-γ and IL-12 receptor β2-chain expression. The production of IFN-γ by the combination of IL-18 plus IL-12 is an example of true synergism in cytokine biology, similar to the synergism of IL-1 and TNF-α in models of inflammation. Since IFN-γ is the "signature" cytokine of CD4+ and CD8+ T cells as well as NK cells, a great deal of the biology of IL-18 is considered due to IFN-γ production. Dendritic cells deficient in the IFN-γ transcription factor T-bet exhibit impaired IFN-γ production after stimulation with IL-18 plus IL-12 [62]. IL-18 is constitutively present in monocytes and monocyte-derived dendritic cells type 1 cells. Thus, IFN-γ induced by the combination of IL-12 plus IL-18 appears to be *via* the T-bet transcription factor.

Graft *versus* host disease

IFN-γ plays a major pathological role in this disease due to its Th1-inducing properties and the generation of cytotoxic T cells. Using a cohort of 157 patients who received unrelated donor bone marrow transplantation and developed graft *versus* host disease, a polymorphism in the IL-18 promoter (G137C, C607A, G656T) was identified and associated with statistically significant decreased risk death [63]. At 100 days after the transplant, the mortality in patients with this polymorphism was 23% compared to 48% in those patients without the polymorphism and after 1 year the mortality was 36% *versus* 65%, respectively. The probability of the survival was twofold in patients with this haplotype [63]. In the case of graft *versus* host disease in mice, paradoxical effects of IL-18 have been reported depending on whether the disease is CD4+ or CD8+ T cell mediated. In humans, T cells are responsible for the disease following allogeneic bone marrow transplantation. Administration of IL-18 to recipient mice increased survival in CD4+-mediated disease but resulted in worsening in the CD8+-mediated disease [64]. Neutralizing anti-IL-18 monoclonal antibodies significantly reduced CD8+-mediated mortality [64]. Administration of IL-18 reduces the severity of the disease by inducing the production of Th2 cytokines [65].

IL-18 and Th2 diseases

The combination of IL-18 plus IL-12 suppresses IgE synthesis *via* IFN-γ production and suggests a role for IL-18 in Th2 polarization. For example, in models of allergic asthma, injecting both IL-12 plus IL-18 suppresses IgE synthesis, eosinophila and airway hyperresponsiveness (reviewed in [7]). In contrast, the administration of IL-18 alone enhanced basophil production of IL-4 and histamine and increased serum IgE levels in wild-type and IL-4-deficient mice [66]. Overexpression of mature IL-18 in the skin results is worsening of allergic and non-allergic cutaneous inflammation *via* Th2 cytokines [67]. Mice overexpressing IL-18 or overexpressing caspase-1 develop an atopic-like dermatitis with mastocytosis and the presence of Th2 cytokines; also present in these mice was elevated serum IgE [68]. Although IL-18 remains a Th1 cytokine, there are increasing reports showing a role for IL-18 in promoting Th2-mediated diseases [69]. Upon neutralization of IL-18 in co-cultures of dendritic cells type 1 cells with allogeneic naive T lymphocytes, the Th1/Th2 phenotype was not affected, whereas anti-IL-12 down-regulated the Th1 response [70]. In fact, IL-18Rs were expressed on dendritic cells of the type-2 lineage, suggesting a Th2 response [70].

IL-18-mediated islet injury

Insulin-producing islet β-cells secrete IL-18 and supernatants from stimulated islets induce IFN-γ in T cells in an IL-18-dependent manner [71]. Inside islets, however, the expression of IL-18R is limited to resident non-β-cells [72]. Islet-derived IL-18 can therefore function by engaging the IL-18R expressed on islet stromal cells, i.e., macrophages, T cells, fibroblasts and endothelial cells [71]. For this reason, the effects of β-cell-derived IL-18 on β-cell responses is observed in intact islets, or in islets surrounded by neighboring cells. Indeed, evidence suggests an association between local IL-18 levels and β-cell damage. Islets isolated from the non-obese diabetic mouse strain exhibit IL-18 expression prior to T cell invasion [71] and exogenous administration of IL-18 worsens diabetes in these mice [73]. IL-18 also contributes to the injury of streptozotocin (STZ)-induced diabetes [74] and IL-18 blockade with IL-18BP delays the development of diabetes in the non-obese diabetic mouse [75]. Similarly, mice deficient in IL-18 exhibit delayed STZ-induced hyperglycemia [76]. In humans, the gene for IL-18 maps to an interval on chromosome 9, where a diabetes susceptibility locus, *Idd2*, resides [77].

IL-18 and the heart

Unexpectedly, IL-18 is an important cytokine in myocardial ischemia reperfusion injury, a model of acute infarctions, where it functions to decrease the contractile force of the heart. It appears that the role of IL-18 in myocardial dysfunction is independent of IFN-γ but likely related to the induction of Fas ligand. Human heart tissue contains preformed IL-18 in macrophages and endothelial cells [28]. Upon reducing IL-18 activity with either IL-18BP or a caspase-1 inhibitor, the functional impairment of the ischemia reperfusion injury was reduced [28]. A neutralizing anti-IL-18 polyclonal antibody resulted in near prevention of endotoxin-induced myocardial suppression in mice, and myocardial IL-1β levels were also reduced [78]. Using caspase-1-deficient mice subjected to ligation of the left anterior descending coronary artery as a model for myocardial infarction, significantly lower mortality was observed in the deficient mice compared to the wild-type mice [79]. Caspase-1-deficient mice also had lower levels of IL-18, metalloproteinase-3 activity and myocyte apoptosis following the injury. In humans, myocardial tissue steady-state levels of IL-18, IL-18Rα chain and IL-18BP mRNA and their respective protein levels were measured in patients with end-stage heart failure. Circulating plasma and myocardial tissue levels of IL-18 were increased in the patients compared to age-match healthy subjects [80]. However, mRNA levels of IL-18 BP were decreased in the failing myocardium. In fact, plasma IL-18 levels were significantly higher in patients who died compared to levels in survivors [80].

There is increasing evidence that IL-18 contributes to atherosclerosis. Unlike the IFN-γ-independent role of IL-18 in ischemic heart disease, the atherosclerotic process involves infiltration of the arterial wall by macrophages and T cells and IFN-γ has been identified in the plaque and considered essential for the disease [81]. Human atherosclerotic plaques from the coronary arteries exhibit increased IL-18 and IL-18Rs compared to non-diseased segments of the same artery [82]. The post-caspase-1 cleavage IL-18 was found to co-localize with macrophages, whereas IL-18Rs were expressed on endothelial and smooth muscle cells. The localization of IL-18 and IL-18Rs in smooth muscle cells is an unexpected but important finding for the pathogenesis of atherosclerosis [81, 82].

Atherosclerotic arterial lesions with infiltrating, lipid-laden macrophages as well as T cells develop spontaneously in male apolipoprotein E (apoE)-deficient mice fed a normal diet. When injected for 30 days with IL-18, these mice exhibited a doubling of the lesion size without a change in serum cholesterol [81]. There was also a fourfold increase in infiltrating T cells. However, when apoE-deficient mice were backcrossed into IFN-γ-deficient mice, the IL-18-induced increase in lesion size was not observed [81]. Although exogenous administration of IL-18 worsened the disease, such an experimental design can be related to the dose of IL-18. Therefore, reduction of natural levels of IL-18 in the apoE-deficient mice is a more rigorous assessment for a role for IL-18 in atherosclerosis. Using apoE-deficient mice and overexpression of IL-18BP by transfection with an IL-18BP-containing plasmid, reduced numbers of infiltrating macrophages and T cells as well as decreases in cell death, and lipid content of the plaques were found [83]. In addition, increases in smooth muscle cells and collagen content suggested a stable plaque phenotype with prevention of progression in this well-established model of human coronary artery disease.

IL-18 and renal ischemia

Like myocardial ischemia-reoxygenation, there is an unexpected role in renal ischemia for IL-18, which is independent of T cells and IFN-γ. Clinically, loss of renal function in patients with septic shock contributes significantly to mortality. IL-18 was measured in patients with acute renal failure as well as patients with poor renal function. There was a remarkably higher level of urinary IL-18 compared to that seen in other renal diseases ($p < 0.001$) [84]. IL-18 was also elevated in the urine of patients with delayed function of cadaveric transplants [84]. The conclusions of the study were that urinary IL-18 is a marker for proximal tubular injury in acute renal failure.

In a large clinical study in intensive care units (ICUs), the level of IL-18 in the urine of patients correlated with the development of renal failure more than creatinine as a predictor of impending renal failure [85]. More impressively, based on IL-18 urine levels, it was possible to predict mortality in the ICU by 48 h, a time period nearly 2 days before other indicators of impending death. These findings in humans are consistent with animal studies. Using a reversible model of acute renal failure, mice deficient in caspase-1 were protected [29], which was due to impaired processing of the IL-18 precursor by caspase-1. Furthermore, wild-type mice were also protected by a preinfusion of neutralizing anti-IL-18 polyclonal antibodies [29]. No protection was afforded by administration of the IL-1 receptor antagonist and, therefore, the model reflects the role of IL-18 rather than IL-1β processing. Although the mechanism for the role of IL-18 in causing acute renal failure remains unclear, it is not related to a decrease in neutrophilic infiltration.

Cardiopulmonary bypass often results in acute renal failure. In 20 patients who developed acute renal failure following bypass surgery, serial urine samples were evaluated for IL-18 levels and compared to 35 matched control patients also undergoing cardiopulmonary bypass but without acute renal injury. Acute renal injury was defined as an increase in serum creatinine of 50% or greater. The findings were remarkable in that elevated creatinine levels occurred 48–72 h following bypass surgery, whereas urine IL-18 was statistically significant increased 4–6 h after the end of surgery [86]. Peak levels of urinary IL-18 were 25-fold at 12 h or more after surgery and remained elevated for 48 h. Multivariate analysis of elevated urinary IL-18 and urinary neutrophil gelatinase-associated lipocalin, also elevated 25-fold early in acute renal failure, revealed that these two markers were independently associated with number of days of acute renal injury. These studies suggest that that elevated urinary IL-18 levels predict acute renal injury after bypass surgery and may be used as a reliable biomarker rather than serum creatinine [86].

IL-18 deficiency triggers overeating, obesity and insulin resistance

Although mice deficient in IL-18 are resistant to various exogenous challenges, an unexpected observation was that as mice aged, they gained significantly more weight than wild-type control mice. By 6 months of age, IL-18-deficient mice were 18.5% heavier than age- and sex-matched wild-type, and by 12 months, 38.1% heavier [87]. The difference in weight was due to more body fat. Basic metabolic rate and core temperature were not different between the two strains but increased food intake accounted for the weight gain. Not unexpectedly, leptin levels were higher in the IL-18-deficient mice and leptin levels correlated with body weight, but there was no evidence that fat mice deficient in IL-18 were resistant to leptin [87]. IL-6 levels were similar in the two groups. The islets of the IL-18-deficient mice exhibited normal architecture but were larger than those of wild-type mice. Histological examination of major organs did not reveal significant difference but the aorta of the IL-18-deficient mice contained lipid deposits characteristic of atherosclerosis [87].

Mice deficient in IL-18 at 6 months of age exhibited elevated fasting glucose compared to wild-type controls, although at 3 months of age, there were no differences between the two groups. Glucose tolerance testing was abnormal in the IL-18-deficient mice and consistent with insulin resistance. Mice deficient in the α chain of the IL-18R also exhibited similar increases in weight at 6 months as well as elevated plasma fasting glucose and insulin resistance. In addition, mice overexpressing the natural inhibitor of IL-18, IL-18BP, overate, gained weight and were hyperglycemic [87]. The administration of recombinant murine IL-18 to the IL-18-deficient mice reversed insulin resistance. The rise in glucose was prevented by the administration of recombinant IL-18 to either the wild-type or the IL-18-deficient mice but not the IL-18R-deficient mice. Others have also reported obese mice lacking IL-IL-18 but that that IL-18 controls energy homeostasis by suppressing appetite and feed efficiency [88].

The mechanism for the increased eating in mice deficient in IL-18, deficient in the IL-18R or in transgenic mice overexpressing the IL-18BP appears to be defect in the control of food intake by the hypothalamic satiety center. Insulin resistance in the liver and muscle were due to the obese condition. Of importance is the observation that, unlike IL-1β, IL-18 does not cause fever [10, 89] and does not induce COX-2 [13]. Phosphorylation of STAT 3 was defective in mice deficient in IL-18. Nevertheless, recombinant IL-18 administered intracerebrally inhibited food intake and, in addition, recombinant IL-18 reversed hyperglycemia in mice deficient for IL-18, through activation of STAT3 phosphorylation [87]. Hepatic genes for glucose neogenesis were increased in mice deficient in IL-18 possibly due to the phosphorylation of STAT 3. In mice deficient for IL-18, there was less constitutive phosphorylation of STAT 3 in the liver. Since IL-18 is constitutively expressed in healthy mice and humans [90], the decrease in STAT 3 phosphorylation may be due to the lack of

IL-18 in these mice. These findings indicate a new role of IL-18 in the homeostasis of energy intake and insulin sensitivity.

Conclusions on the therapeutic targeting of IL-18

Exploiting discoveries such as IL-18 to improve the treatment of disease depends on the validity of preclinical research, the resources of the developer and the influence of market forces. There are no doubts in our opinion that IL-18 plays a role in several diseases but despite its use in murine models of tumors, the administration of IL-18 as a therapeutic in human cancer remains unlikely. The challenge for a cytokine like IL-18 is therefore which diseases to target and which agents are best to reduce IL-18 activities. One disease that may benefit blocking IL-18 is MAS. Current treatment of life-threatening MAS is intravenous cyclosporine A, a nephrotoxic inhibitor of IFN-γ and other T cell cytokines and high doses of corticosteroids. The role of IL-18 in IFN-γ production and the role of IFN-γ in macrophage activation are well established. The finding that patients with active MAS have high circulating levels of free IL-18 due to lower than expected levels of IL-18BP provides a rational basis for reducing IL-18 in MAS [91]. To prove this concept would require treating patients with progressive or established MAS with increasing doses of IL-18BP and monitoring ferritin levels as well as clinical responses. Because MAS can be a fatal disease, testing the concept requires no placebo arm and few patients. Outcomes of clinical improvements, a reduction in corticosteroids and weaning of cyclosporine A, would be sufficient for orphan drug status approval. Oral caspase-1 inhibitors also deserve testing in patients with MAS.

It is also our opinion that another acute and life-threatening disease can be subjected to a proof of concept, e.g., acute renal failure. Caspase-1 inhibitors would require intravenous administration rather than the oral route. However, preclinical testing in mice deficient in caspase-1 provides a rationale for the testing of caspase-1 inhibitors in patients at risk for acute renal failure. Since IL-18BP can be readily administered subcutaneously or intravenously, IL-18BP could be used in a trial and compared to a placebo arm. There may be reluctance to test a potentially useful anti-cytokine in acute renal failure, since this disease is essentially the consequence of septic shock. The disappointing results of TNF and IL-1 blockers in sepsis trials may reduce enthusiasm for treating these patients. However, one important benefit for testing IL-18BP in preventing or reducing acute renal failure is the predictive value of urinary IL-18 levels. Since urinary IL-18 is elevated in patients at risk a full 48 h before the renal dysfunction becomes apparent [84], the design of a trial of IL-18BP infusion would be restricted to a subgroup of those patients. The clinical application of urinary IL-18 determinations as an entry criterion is substantial due to its quantification compared to clinical scoring methods. A reduction in progression to renal failure compared to placebo-treated patients would provide a basis for approval since preventing acute renal failure is an unmet medical need.

The third clinical setting for testing agents that reduce IL-18 activity is acute lupus nephritis and vasculitis. Here the role of IL-18 in the production of IFN-γ may be of paramount importance. Animal models of lupus indicate a pathological role for IL-18 in the kidney. However, the effect of IL-18 on the vasculature may also be part of the lupus vasculitis. Some early trials have blocked TNF-α in acute lupus nephritis using monoclonal anti-TNF-α antibodies. In this case, these antibodies reduce both TNF-α and IFN-γ. Taken together, these preclinical observations provide a rationale for an intervention study in lupus nephritis. Testing the concept with oral caspase-1 inhibitors or IL-18BP administered subcutaneously asks whether the vessel wall inflammation can be reduced over that presently achieved by heparin, corticosteroids and aspirin.

Acknowledgements

These studies are supported by Supported by NIH Grants AI-15614 and HL-68743 and the Colorado Cancer Center.

References

1   Okamura H, Tsutsui H, Komatsu T, Yutsudo M, Hakura A, Tanimoto T, Torigoe K, Okura T, Nukada Y, Hattori K et al (1995) Cloning of a new cytokine that induces interferon-γ. *Nature* 378: 88–91

2   Schmitz J, Owyang A, Oldham E, Song Y, Murphy E, McClanahan TK, Zurawski G, Moshrefi M, Qin J, Li X et al (2005) IL-33, an interleukin-1-like cytokine that signals *via* the IL-1 receptor-related protein ST2 and induces T helper type 2-associated cytokines. *Immunity* 23: 479–490

3   Ghayur T, Banerjee S, Hugunin M, Butler D, Herzog L, Carter A, Quintal L, Sekut L, Talanian R, Paskind M et al (1997) Caspase-1 processes IFN-gamma-inducing factor and regulates LPS-induced IFN-gamma production. *Nature* 386: 619–623

4   Gu Y, Kuida K, Tsutsui H, Ku G, Hsiao K, Fleming MA, Hayashi N, Higashino K, Okamura H, Nakanishi K et al (1997) Activation of interferon-γ inducing factor mediated by interleukin-1β converting enzyme. *Science* 275: 206–209

5   Mariathasan S, Newton K, Monack DM, Vucic D, French DM, Lee WP, Roose-Girma M, Erickson S, Dixit VM (2004) Differential activation of the inflammasome by caspase-1 adaptors ASC and Ipaf. *Nature* 430: 213–218

6   Fantuzzi G, Puren AJ, Harding MW, Livingston DJ, Dinarello CA (1998) IL-18 regulation of IFN-γ production and cell proliferation as revealed in interleukin-1β converting enzyme-deficient mice. *Blood* 91: 2118–2125

7   Nakanishi K, Yoshimoto T, Tsutsui H, Okamura H (2001) Interleukin-18 is a unique cytokine that stimulates both Th1 and Th2 responses depending on its cytokine milieu. *Cytokine Growth Factor Rev* 12: 53–72

8   Siegmund B, Lehr HA, Fantuzzi G, Dinarello CA (2001) IL-1beta -converting enzyme (caspase-1) in intestinal inflammation. *Proc Natl Acad Sci USA* 98: 13249–13254

9   Siegmund B, Fantuzzi G, Rieder F, Gamboni-Robertson F, Lehr HA, Hartmann G, Dinarello CA, Endres S, Eigler A (2001) Neutralization of interleukin-18 reduces severity in murine colitis and intestinal IFN-γ and TNF-α production. *Am J Physiol Regul Integr Comp Physiol* 281: R1264–1273

10 Gatti S, Beck J, Fantuzzi G, Bartfai T, Dinarello CA (2002) Effect of interleukin-18 on mouse core body temperature. *Am J Physiol Regul Integr Comp Physiol* 282: R702–709

11 Li S, Goorha S, Ballou LR, Blatteis CM (2003) Intracerebroventricular interleukin-6, macrophage inflammatory protein-1 beta and IL-18: Pyrogenic and PGE(2)-mediated? *Brain Res* 992: 76–84

12 Reznikov LL, Kim SH, Westcott JY, Frishman J, Fantuzzi G, Novick D, Rubinstein M, Dinarello CA (2000) IL-18 binding protein increases spontaneous and IL-1-induced prostaglandin production *via* inhibition of IFN-gamma. *Proc Natl Acad Sci USA* 97: 2174–2179

13 Lee JK, Kim SH, Lewis EC, Azam T, Reznikov LL, Dinarello CA (2004) Differences in signaling pathways by IL-1beta and IL-18. *Proc Natl Acad Sci USA* 101: 8815–8820

14 Kim SH, Han SY, Azam T, Yoon DY, Dinarello CA (2005) Interleukin-32: A cytokine and inducer of TNFalpha. *Immunity* 22: 131–142

15 Joosten LA, van De Loo FA, Lubberts E, Helsen MM, Netea MG, van Der Meer JW, Dinarello CA, van Den Berg WB (2000) An IFN-gamma-independent proinflammatory role of IL-18 in murine streptococcal cell wall arthritis. *J Immunol* 165: 6553–6558

16 Carrascal MT, Mendoza L, Valcarcel M, Salado C, Egilegor E, Telleria N, Vidal-Vanaclocha F, Dinarello CA (2003) Interleukin-18 binding protein reduces b16 melanoma hepatic metastasis by neutralizing adhesiveness and growth factors of sinusoidal endothelium. *Cancer Res* 63: 491–497

17 Randle JC, Harding MW, Ku G, Schonharting M, Kurrle R (2001) ICE/caspase-1 inhibitors as novel anti-inflammatory drugs. *Expert Opin Investig Drugs* 10: 1207–1209

18 Novick D, Kim S-H, Fantuzzi G, Reznikov L, Dinarello CA, Rubinstein M (1999) Interleukin-18 binding protein: A novel modulator of the Th1 cytokine response. *Immunity* 10: 127–136

19 Novick D, Schwartsburd B, Pinkus R, Suissa D, Belzer I, Sthoeger Z, Keane WF, Chvatchko Y, Kim SH, Fantuzzi G et al (2001) A novel IL-18BP ELISA shows elevated serum il-18BP in sepsis and extensive decrease of free IL-18. *Cytokine* 14: 334–342

20 Kim S-H, Eisenstein M, Reznikov L, Fantuzzi G, Novick D, Rubinstein M, Dinarello CA (2000) Structural requirements of six naturally occurring isoforms of the interleukin-18 binding protein to inhibit interleukin-18. *Proc Natl Acad Sci USA* 97: 1190–1195

21 Tak PP, Bacchi M, Bertolino M (2006) Pharmacokinetics of IL-18 binding protein in

healthy volunteers and subjects with rheumatoid arthritis or plaque psoriasis. *Eur J Drug Metab Pharmacokinetics* 31: 109–116

22 Agostini L, Martinon F, Burns K, McDermott MF, Hawkins PN, Tschopp J (2004) NALP3 forms an IL-1β processing inflammasome with increased activity in Muckle-Wells auto-inflammatory disorder. *Immunity* 20: 319–325

23 Aganna E, Martinon F, Hawkins PN, Ross JB, Swan DC, Booth DR, Lachmann HJ, Bybee A, Gaudet R, Woo P et al (2002) Association of mutations in the NALP3/CIAS1/PYPAF1 gene with a broad phenotype including recurrent fever, cold sensitivity, sensorineural deafness, and AA amyloidosis. *Arthritis Rheum* 46: 2445–2452

24 Hawkins PN, Lachmann HJ, Aganna E, McDermott MF (2004) Spectrum of clinical features in Muckle-Wells syndrome and response to anakinra. *Arthritis Rheum* 50: 607–612

25 Coeshott C, Ohnemus C, Pilyavskaya A, Ross S, Wieczorek M, Kroona H, Leimer AH, Cheronis J (1999) Converting enzyme-independent release of TNFα and IL-1β from a stimulated human monocytic cell line in the presence of activated neutrophils or purified proteinase-3. *Proc Natl Acad Sci USA* 96: 6261–6266

26 Sugawara S, Uehara A, Nochi T, Yamaguchi T, Ueda H, Sugiyama A, Hanzawa K, Kumagai K, Okamura H, Takada H (2001) Neutrophil proteinase 3-mediated induction of bioactive IL-18 secretion by human oral epithelial cells. *J Immunol* 167: 6568–6575

27 Faggioni R, Cattley RC, Guo J, Flores S, Brown H, Qi M, Yin S, Hill D, Scully S, Chen C et al (2001) IL-18-binding protein protects against lipopolysaccharide-induced lethality and prevents the development of Fas/Fas ligand-mediated models of liver disease in mice. *J Immunol* 167: 5913–5920

28 Pomerantz BJ, Reznikov LL, Harken AH, Dinarello CA (2001) Inhibition of caspase 1 reduces human myocardial ischemic dysfunction *via* inhibition of IL-18 and IL-1beta. *Proc Natl Acad Sci USA* 98: 2871–2876

29 Melnikov VY, Ecder T, Fantuzzi G, Siegmund B, Lucia MS, Dinarello CA, Schrier RW, Edelstein CL (2001) Impaired IL-18 processing protects caspase-1–deficient mice from ischemic acute renal failure. *J Clin Invest* 107: 1145–1152

30 Perregaux DG, McNiff P, Laliberte R, Conklyn M, Gabel CA (2000) ATP acts as an agonist to promote stimulus-induced secretion of IL-1 beta and IL-18 in human blood. *J Immunol* 165: 4615–4623

31 Solle M, Labasi J, Perregaux DG, Stam E, Petrushova N, Koller BH, Griffiths RJ, Gabel CA (2001) Altered cytokine production in mice lacking P2X(7) receptors. *J Biol Chem* 276: 125–132

32 Laliberte RE, Eggler J, Gabel CA (1999) ATP treatment of human monocytes promotes caspase-1 maturation and externalization. *J Biol Chem* 274: 36944–36951

33 Baraldi PG, del Carmen Nunez M, Morelli A, Falzoni S, Di Virgilio F, Romagnoli R (2003) Synthesis and biological activity of *N*-arylpiperazine-modified analogues of KN-62, a potent antagonist of the purinergic P2X7 receptor. *J Med Chem* 46: 1318–1329

34 Gardella S, Andrei C, Costigliolo S, Poggi A, Zocchi MR, Rubartelli A (1999) Inter-

leukin-18 synthesis and secretion by dendritic cells are modulated by interaction with antigen-specific T cells. *J Leukoc Biol* 66: 237–241

35 Gardella S, Andrei C, Poggi A, Zocchi MR, Rubartelli A (2000) Control of interleukin-18 secretion by dendritic cells: Role of calcium influxes. *FEBS Lett* 481: 245–248

36 Kato Z, Jee J, Shikano H, Mishima M, Ohki I, Ohnishi H, Li A, Hashimoto K, Matsukuma E, Omoya K et al (2003) The structure and binding mode of interleukin-18. *Nat Struct Biol* 10: 966–971

37 Azam T, Novick D, Bufler P, Yoon DY, Rubinstein M, Dinarello CA, Kim SH (2003) Identification of a critical Ig-like domain in IL-18 receptor alpha and characterization of a functional IL-18 receptor complex. *J Immunol* 171: 6574–6580

38 Casadio R, Frigimelica E, Bossu P, Neumann D, Martin MU, Tagliabue A, Boraschi D (2001) Model of interaction of the IL-1 receptor accessory protein IL-1RAcP with the IL-1beta/IL-1R(I) complex. *FEBS Lett* 499: 65–68

39 Reznikov LL, Kim SH, Zhou L, Bufler P, Goncharov I, Tsang M, Dinarello CA (2002) The combination of soluble IL-18Rα and IL-18Rβ chains inhibits IL-18-induced IFN-γ. *J Interferon Cytokine Res* 22: 593–601

40 Kumar S, McDonnell PC, Lehr R, Tierney L, Tzimas MN, Griswold DE, Capper EA, Tal-Singer R, Wells GI, Doyle ML et al (2000) Identification and initial characterization of four novel members of the interleukin-1 family. *J Biol Chem* 275: 10308–10314

41 Pan G, Risser P, Mao W, Baldwin DT, Zhong AW, Filvaroff E, Yansura D, Lewis L, Eigenbrot C, Henzel WJ et al (2001) IL-1H, an interleukin 1-related protein that binds IL-18 receptor/IL-1Rrp. *Cytokine* 13: 1–7

42 Kumar S, Hanning CR, Brigham-Burke MR, Rieman DJ, Lehr R, Khandekar S, Kirkpatrick RB, Scott GF, Lee JC, Lynch FJ et al (2002) Interleukin-1F7B (IL-1H4/IL-1F7) is processed by caspase-1 and mature IL-1F7B binds to the IL-18 receptor but does not induce IFN-gamma production. *Cytokine* 18: 61–71

43 Bufler P, Azam T, Gamboni-Robertson F, Reznikov LL, Kumar S, Dinarello CA, Kim SH (2002) A complex of the IL-1 homologue IL-1F7b and IL-18–binding protein reduces IL-18 activity. *Proc Natl Acad Sci USA* 99: 13723–13728

44 Novick D, Engelmann H, Wallach D, Leitner O, Revel M, Rubinstein M (1990) Purification of soluble cytokine receptors from normal human urine by ligand-affinity and immunoaffinity chromatography. *J Chromatogr* 510: 331–337

45 Engelmann H, Novick D, Wallach D (1990) Two tumor necrosis factor-binding proteins purified from human urine. Evidence for immunological cross-reactivity with cell surface tumor necrosis factor receptors. *J Biol Chem* 265: 1531–1536

46 Kim SH, Azam T, Novick D, Yoon DY, Reznikov LL, Bufler P, Rubinstein M, Dinarello CA (2002) Identification of amino acid residues critical for biological activity in human interleukin-18. *J Biol Chem* 277: 10998–11003

47 Dinarello CA (2005) The many worlds of reducing interleukin-1. *Arthritis Rheum* 52: 1960–1967

48 Hurgin V, Novick D, Rubinstein M (2002) The promoter of IL-18 binding protein:

Activation by an IFN-gamma-induced complex of IFN regulatory factor 1 and CCAAT/ enhancer binding protein beta. *Proc Natl Acad Sci USA* 99: 16957–16962

49   Paulukat J, Bosmann M, Nold M, Garkisch S, Kampfer H, Frank S, Raedle J, Zeuzem S, Pfeilschifter J, Muhl H (2001) Expression and release of IL-18 binding protein in response to IFNγ. *J Immunol* 167: 7038–7043

50   Fantuzzi G, Reed D, Qi M, Scully S, Dinarello CA, Senaldi G (2001) Role of interferon regulatory factor-1 in the regulation of IL-18 production and activity. *Eur J Immunol* 31: 369–375

51   Siegmund B, Sennello JA, Lehr HA, Senaldi G, Dinarello CA, Fantuzzi G (2004) Interferon regulatory factor-1 as a protective gene in intestinal inflammation: Role of TCR gamma delta T cells and interleukin-18-binding protein. *Eur J Immunol* 34: 2356–2364

52   Xiang Y, Moss B (2001) Correspondence of the functional epitopes of poxvirus and human interleukin-18–binding proteins. *J Virol* 75: 9947–9954

53   Xiang Y, Moss B (2001) Determination of the functional epitopes of human interleukin-18–binding protein by site-directed mutagenesis. *J Biol Chem* 276: 17380–17386

54   Maeno N, Takei S, Imanaka H, Yamamoto K, Kuriwaki K, Kawano Y, Oda H (2004) Increased interleukin-18 expression in bone marrow of a patient with systemic juvenile idiopathic arthritis and unrecognized macrophage-activation syndrome. *Arthritis Rheum* 50: 1935–1938

55   Dinarello CA (2005) Differences between anti-tumor necrosis factor-alpha monoclonal antibodies and soluble TNF receptors in host defense impairment. *J Rheumatol Suppl* 74: 40–47

56   Ottenhoff TH, Verreck FA, Lichtenauer-Kaligis EG, Hoeve MA, Sanal O, van Dissel JT (2002) Genetics, cytokines and human infectious disease: Lessons from weakly pathogenic mycobacteria and salmonellae. *Nat Genet* 32: 97–105

57   Wei XQ, Leung BP, Arthur HM, McInnes IB, Liew FY (2001) Reduced incidence and severity of collagen-induced arthritis in mice lacking IL-18. *J Immunol* 166: 517–521

58   Plater-Zyberk C, Joosten LA, Helsen MM, Sattonnet-Roche P, Siegfried C, Alouani S, van De Loo FA, Graber P, Aloni S, Cirillo R et al (2001) Therapeutic effect of neutralizing endogenous IL-18 activity in the collagen-induced model of arthritis. *J Clin Invest* 108: 1825–1832

59   Bossu P, Neumann D, Del Giudice E, Ciaramella A, Gloaguen I, Fantuzzi G, Dinarello CA, Di Carlo E, Musiani P, Meroni PL et al (2003) IL-18 cDNA vaccination protects mice from spontaneous lupus-like autoimmune disease. *Proc Natl Acad Sci USA* 100: 14181–14186

60   Kim SH, Reznikov LL, Stuyt RJ, Selzman CH, Fantuzzi G, Hoshino T, Young HA, Dinarello CA (2001) Functional reconstitution and regulation of IL-18 activity by the IL-18R beta chain. *J Immunol* 166: 148–154

61   Neumann D, Martin MU (2001) Interleukin-12 upregulates the IL-18Rβ chain in BALB/c thymocytes. *J Interferon Cytokine Res* 21: 635–642

62   Lugo-Villarino G, Maldonado-Lopez R, Possemato R, Penaranda C, Glimcher LH

(2003) T-bet is required for optimal production of IFN-gamma and antigen-specific T cell activation by dendritic cells. *Proc Natl Acad Sci USA* 100: 7749–7754

63 Cardoso SM, DeFor TE, Tilley LA, Bidwell JL, Weisdorf DJ, MacMillan ML (2004) Patient interleukin-18 GCG haplotype associates with improved survival and decreased transplant-related mortality after unrelated-donor bone marrow transplantation. *Br J Haematol* 126: 704–710

64 Min CK, Maeda Y, Lowler K, Liu C, Clouthier S, Lofthus D, Weisiger E, Ferrara JL, Reddy P (2004) Paradoxical effects of interleukin-18 on the severity of acute graft-*versus*-host disease mediated by CD4+ and CD8+ T-cell subsets after experimental allogeneic bone marrow transplantation. *Blood* 104: 3393–3399

65 Reddy P, Ferrara JL (2003) Role of interleukin-18 in acute graft-vs-host disease. *J Lab Clin Med* 141: 365–371

66 Hoshino T, Yagita H, Ortaldo JR, Wiltrout RH, Young HA (2000) *In vivo* administration of IL-18 can induce IgE production through Th2 cytokine induction and up-regulation of CD40 ligand (CD154) expression on CD4+ T cells. *Eur J Immunol* 30: 1998–2006

67 Kawase Y, Hoshino T, Yokota K, Kuzuhara A, Kirii Y, Nishiwaki E, Maeda Y, Takeda J, Okamoto M, Kato S et al (2003) Exacerbated and prolonged allergic and non-allergic inflammatory cutaneous reaction in mice with targeted interleukin-18 expression in the skin. *J Invest Dermatol* 121: 502–509

68 Konishi H, Tsutsui H, Murakami T, Yumikura-Futatsugi S, Yamanaka K, Tanaka M, Iwakura Y, Suzuki N, Takeda K, Akira S et al (2002) IL-18 contributes to the spontaneous development of atopic dermatitis-like inflammatory skin lesion independently of IgE/stat6 under specific pathogen-free conditions. *Proc Natl Acad Sci USA* 99: 11340–11345

69 Nakanishi K, Yoshimoto T, Tsutsui H, Okamura H (2001) Interleukin-18 regulates both Th1 and Th2 responses. *Ann Rev Immunol* 19: 423–474

70 Kaser A, Kaser S, Kaneider NC, Enrich B, Wiedermann CJ, Tilg H (2004) Interleukin-18 attracts plasmacytoid dendritic cells (DC2s) and promotes Th1 induction by DC2s through IL-18 receptor expression. *Blood* 103: 648–655

71 Frigerio S, Hollander GA, Zumsteg U (2002) Functional IL-18 Is produced by primary pancreatic mouse islets and NIT-1 beta cells and participates in the progression towards destructive insulitis. *Horm Res* 57: 94–104

72 Hong TP, Andersen NA, Nielsen K, Karlsen AE, Fantuzzi G, Eizirik DL, Dinarello CA, Mandrup-Poulsen T (2000) Interleukin-18 mRNA, but not interleukin-18 receptor mRNA, is constitutively expressed in islet beta-cells and up-regulated by interferon-gamma. *Eur Cytokine Netw* 11: 193–205

73 Oikawa Y, Shimada A, Kasuga A, Morimoto J, Osaki T, Tahara H, Miyazaki T, Tashiro F, Yamato E, Miyazaki J et al (2003) Systemic administration of IL-18 promotes diabetes development in young nonobese diabetic mice. *J Immunol* 171: 5865–5875

74 Nicoletti F, Di Marco R, Papaccio G, Conget I, Gomis R, Bernardini R, Sims JE, Shoenfeld Y, Bendtzen K (2003) Essential pathogenic role of endogenous IL-18 in murine

diabetes induced by multiple low doses of streptozotocin. Prevention of hyperglycemia and insulitis by a recombinant IL-18–binding protein: Fc construct. *Eur J Immunol* 33: 2278–2286

75 Zaccone P, Phillips J, Conget I, Cooke A, Nicoletti F (2005) IL-18 binding protein fusion construct delays the development of diabetes in adoptive transfer and cyclophosph-amide-induced diabetes in NOD mouse. *Clin Immunol* 115: 74–79

76 Lukic ML, Mensah-Brown E, Wei X, Shahin A, Liew FY (2003) Lack of the mediators of innate immunity attenuate the development of autoimmune diabetes in mice. *J Auto-immun* 21: 239–246

77 Sarvetnick N (1997) IFN-gamma, IGIF, and IDDM. *J Clin Invest* 99: 371–372

78 Raeburn CD, Dinarello CA, Zimmerman MA, Calkins CM, Pomerantz BJ, McIntyre RC Jr, Harken AH, Meng X (2002) Neutralization of IL-18 attenuates lipopolysaccha-ride-induced myocardial dysfunction. *Am J Physiol* 283: H650–657

79 Friteau L, Francesconi E, Lando D, Dugas B, Damais C (1988) Opposite effect of interferon-γ on PGE2 release from interleukin-1-stimulated human monocytes or fibro-blasts. *Biochem Biophys Res Commun* 157: 1197–1204

80 Mallat Z, Heymes C, Corbaz A, Logeart D, Alouani S, Cohen-Solal A, Seidler T, Hasen-fuss G, Chvatchko Y, Shah AM et al (2004) Evidence for altered interleukin 18 (IL)-18 pathway in human heart failure. *FASEB J* 18: 1752–1754

81 Whitman SC, Ravisankar P, Daugherty A (2002) Interleukin-18 enhances atheroscle-rosis in apolipoprotein E(–/–) mice through release of interferon-gamma. *Circ Res* 90: E34–38

82 Gerdes N, Sukhova GK, Libby P, Reynolds RS, Young JL, Schonbeck U (2002) Expres-sion of interleukin (IL)-18 and functional IL-18 receptor on human vascular endothelial cells, smooth muscle cells, and macrophages: Implications for atherogenesis. *J Exp Med* 195: 245–257

83 Mallat Z, Corbaz A, Scoazec A, Graber P, Alouani S, Esposito B, Humbert Y, Chvatchko Y, Tedgui A (2001) Interleukin-18/interleukin-18 binding protein signaling modulates atherosclerotic lesion development and stability. *Circ Res* 89: E41–45

84 Parikh CR, Jani A, Melnikov VY, Faubel S, Edelstein CL (2004) Urinary interleukin-18 is a marker of human acute tubular necrosis. *Am J Kidney Dis* 43: 405–414

85 Parikh CR, Abraham E, Ancukiewicz M, Edelstein CL (2005) Urine IL-18 is an early diagnostic marker for acute kidney injury and predicts mortality in the ICU. *J Am Soc Nephrol* 16: 3046–3052

86 Parikh CR, Mishra J, Thiessen-Philbrook H, Dursun B, Ma Q, Kelly C, Dent C, Devara-jan P, Edelstein CL (2006) Urinary IL-18 is an early predictive biomarker of acute kidney injury after cardiac surgery. *Kidney Int* 70: 199–203

87 Netea MG, Joosten LA, Lewis E, Jensen DR, Voshol PJ, Kullberg BJ, Tack CJ, van Krieken H, Kim SH, Stalenhoef AF et al (2006) Deficiency of interleukin-18 in mice leads to hyperphagia, obesity and insulin resistance. *Nat Med* 12: 650–656

88 Zorrilla EP, Sanchez-Alavez M, Sugama S, Brennan M, Fernandez R, Bartfai T, Conti

B (2007) Interleukin-18 controls energy homeostasis by suppressing appetite and feed efficiency. *Proc Natl Acad Sci USA* 104: 11097–11102

89   Stuyt RJ, Netea MG, Verschueren I, Dinarello CA, Kullberg BJ, van der Meer JW (2005) Interleukin-18 does not modulate the acute-phase response. *J Endotoxin Res* 11: 85–88

90   Puren AJ, Fantuzzi G, Dinarello CA (1999) Gene expression, synthesis and secretion of IL-1β and IL-18 are differentially regulated in human blood mononuclear cells and mouse spleen cells. *Proc Natl Acad Sci USA* 96: 2256–2261

91   Mazodier K, Marin V, Novick D, Farnarier C, Robitail S, Schleinitz N, Veit V, Paul P, Rubinstein M, Dinarello CA et al (2005) Severe imbalance of IL-18/IL-18BP in patients with secondary hemophagocytic syndrome. *Blood* 106: 3483–3489

92   Sivakumar PV, Westrich GM, Kanaly S, Garka K, Born TL, Derry JM, Viney JL (2002) Interleukin 18 is a primary mediator of the inflammation associated with dextran sulphate sodium induced colitis: Blocking interleukin 18 attenuates intestinal damage. *Gut* 50: 812–820

93   Ten Hove T, Corbaz A, Amitai H, Aloni S, Belzer I, Graber P, Drillenburg P, van Deventer SJ, Chvatchko Y, Te Velde AA (2001) Blockade of endogenous IL-18 ameliorates TNBS-induced colitis by decreasing local TNF-alpha production in mice. *Gastroenterology* 121: 1372–1379

94   Wirtz S, Becker C, Blumberg R, Galle PR, Neurath MF (2002) Treatment of T cell-dependent experimental colitis in SCID mice by local administration of an adenovirus expressing IL-18 antisense mRNA. *J Immunol* 168: 411–420

95   Smeets RL, van de Loo FA, Arntz OJ, Bennink MB, Joosten LA, van den Berg WB (2003) Adenoviral delivery of IL-18 binding protein C ameliorates collagen-induced arthritis in mice. *Gene Ther* 10: 1004–1011

96   Maecker HT, Hansen G, Walter DM, DeKruyff RH, Levy S, Umetsu DT (2001) Vaccination with allergen-IL-18 fusion DNA protects against, and reverses established, airway hyperreactivity in a murine asthma model. *J Immunol* 166: 959–965

97   Im SH, Barchan D, Maiti PK, Raveh L, Souroujon MC, Fuchs S (2001) Suppression of experimental myasthenia gravis, a B cell-mediated autoimmune disease, by blockade of IL-18. *FASEB J* 15: 2140–2148

98   Furlan R, Martino G, Galbiati F, Poliani PL, Smiroldo S, Bergami A, Desina G, Comi G, Flavell R, Su MS et al (1999) Caspase-1 regulates the inflammatory process leading to autoimmune demyelination. *J Immunol* 163: 2403–2409

99   Faggioni R, Jones-Carson J, Reed DA, Dinarello CA, Feingold KR, Grunfeld C, Fantuzzi G (2000) Leptin-deficient (ob/ob) mice are protected from T cell-mediated hepatotoxicity: Role of tumor necrosis factor alpha and IL-18. *Proc Natl Acad Sci USA* 97: 2367–2372

100   Fantuzzi G, Banda NK, Guthridge C, Vondracek A, Kim SH, Siegmund B, Azam T, Sennello JA, Dinarello CA, Arend WP (2003) Generation and characterization of mice transgenic for human IL-18-binding protein isoform a. *J Leukoc Biol* 74: 889–896

101   Tsutsui H, Kayagaki N, Kuida K, Nakano H, Hayashi N, Takeda K, Matsui K, Kashi-

wamura S, Hada T, Akira S et al (1999) Caspase-1-independent, Fas/Fas ligand-mediated IL-18 secretion from macrophages causes acute liver injury in mice. *Immunity* 11: 359–367

102 Fantuzzi G, Reed DA, Dinarello CA (1999) IL-12-induced IFNγ is dependent on caspase-1 processing of the IL-18 precursor. *J Clin Invest* 104: 761–767

103 Netea MG, Fantuzzi G, Kullberg BJ, Stuyt RJ, Pulido EJ, McIntyre RC Jr, Joosten LA, Van der Meer JW, Dinarello CA (2000) Neutralization of IL-18 reduces neutrophil tissue accumulation and protects mice against lethal *Escherichia coli* and *Salmonella typhimurium* endotoxemia. *J Immunol* 164: 2644–2649

104 Vidal-Vanaclocha F, Fantuzzi G, Mendoza L, Fuentes AM, Anasagasti MJ, Martin J, Carrascal T, Walsh P, Reznikov LL, Kim SH et al (2000) IL-18 regulates IL-1beta-dependent hepatic melanoma metastasis *via* vascular cell adhesion molecule-1. *Proc Natl Acad Sci USA* 97: 734–739

105 Takeuchi D, Yoshidome H, Kato A, Ito H, Kimura F, Shimizu H, Ohtsuka M, Morita Y, Miyazaki M (2004) Interleukin 18 causes hepatic ischemia/reperfusion injury by suppressing anti-inflammatory cytokine expression in mice. *Hepatology* 39: 699–710

Chemokines and chemokine receptors

Zoltán Szekanecz[1] and Alisa E. Koch[2,3]

[1]Department of Rheumatology, Institute of Medicine, University of Debrecen Medical and Health Science Center, Debrecen, 4004, Hungary
[2]Veterans' Administration, Ann Arbor Healthcare System, Ann Arbor, MI, USA
[3]University of Michigan Health System, Department of Internal Medicine, Division of Rheumatology, Ann Arbor, MI, USA

Abstract

There is a structural and a functional classification of chemokines. The former includes four groups: CXC, CC, C and CX_3C chemokines. There is a redundancy and binding promiscuity between chemokine receptors and their ligands. Recently, a functional classification distinguishing between inflammatory and homeostatic chemokines has been introduced. However, numerous effects of these chemokines overlap. For example, numerous homeostatic chemokines, which are involved in lymphocyte recruitment and lymphoid tissue organization, may also play a role in B cell migration underlying germinal center formation within the inflamed synovium. Anti-chemokine and anti-chemokine receptor targeting may be therapeutically used in future biological therapy of arthritis. In addition to the clear clinical benefit, we can learn a lot from these trials about the actions of the targeted chemokines and their receptors. Today, most data in this field are obtained from experimental models of arthritis; however, results of some human trials have also become available. Thus, it is possible that a number of specific chemokine and chemokine receptor antagonists will be administered to arthritis patients in the near future. Hopefully, some of these potential treatment modalities will be used to control inflammation, prevent joint destruction and thus will benefit our patients.

Introduction

Chemokines are mediators exerting chemotactic activity towards leukocytes under inflammatory conditions, such as rheumatoid arthritis (RA). Target cells express chemokine receptors. There are two major classification systems, a structural and a functional one. Chemokines have been classified into four supergene families with respect to their structure depending on the position of cysteine (C) residues (for reviews see [1–11]). Chemokines have also been functionally categorized as being homeostatic or inflammatory. Homeostatic chemokines are constitutively expressed on cells and are involved in housekeeping functions including the organization of lymphoid structures. In contrast, inflammatory chemokines are inducible upon cytokine activation and are mostly involved in cell recruitment during inflamma-

New Therapeutic Targets in Rheumatoid Arthritis, edited by Paul-Peter Tak
© 2009 Birkhäuser Verlag Basel/Switzerland

tion. These functions often overlap: for example, as described below, some homeostatic chemokines have been implicated in the pathogenesis of RA. Many of these chemokines are also angiogenic or angiostatic (for reviews see [5, 12, 13]).

First we give a brief overview of the chemokine and chemokine receptor families. The inflammatory, angiogenic/angiostatic and homeostatic chemokines and chemokine receptors that are involved in the pathogenesis of RA and thus may become targets for anti-chemokine therapy are discussed in more detail. We also summarize recent targeting data obtained in RA trials and in animal models of arthritis. It is very likely that several anti-chemokine and anti-chemokine receptor trials will be conducted in RA during the next decade.

Chemokines in RA

In RA, chemokines drive inflammatory leukocytes into the inflamed synovial tissue (ST) (for reviews see [1–8, 14, 15]). As mentioned above, chemokines have been classified into four distinct supergene families designated as CXC, CC, C and CX_3C chemokines (Tab. 1). The respective receptor types of these chemokine subsets are CXCR, CCR, CR and CX_3CR [1–3, 6–8, 16] (Tab. 1). There are about 50 known chemokines and 19 chemokine receptors (for reviews see [1–9]) (Tab. 1). There are two nomenclatures for chemokines: apart from their unique classical name (see later) they are also considered as chemokine ligands and they have been assigned a designation of CXCL(1–16), CCL(1–28), XCL(1, 2) or CX_3CL1 (1, 4, 8, 16) (Tab. 1). In this review, both designations are used.

CXC chemokines

Most CXC chemokines implicated in RA chemoattract neutrophils, however, platelet factor 4 (PF4)/CXCL4 and interferon-γ (IFN-γ)-inducible 10-kDa protein (IP-10)/CXCL10 rather drive mononuclear cells into the synovium [7, 8]. CXC chemokines exert several other actions during inflammation: they may stimulate leukocyte adhesion and integrin expression, cytoskeletal reorganization, neutrophil degranulation, respiratory burst and phagocytosis, as well as the production of proteolytic enzymes, prostanoids and platelet-activating factor (for reviews see [7]). In addition, some CXC chemokines induce, while others, suppress synovial angiogenesis [3, 17]. In general, CXC chemokines containing the ELR amino acid residue, such as interleukin-8 (IL-8)/CXCL8, epithelial-neutrophil activating protein-78 (ENA-78)/CXCL5, growth-regulated oncogene α (groα)/CXCL1, and connective tissue activating peptide-III (CTAP-III)/CXCL7 stimulate neovascularization. ELR$^-$ chemokines, such as PF4/CXCL4, IP-10/CXCL10 and monokine induced by IFN-γ (Mig)/CXCL9 are angiostatic [3, 17].

Table 1 - Chemokines and chemokine receptors implicated in rheumatoid arthritis. *

Chemokine receptor	Chemokine ligand
CXC chemokine receptors	
CXCR1	IL-8/CXCL8, GCP-2/CXCL6
CXCR2	IL-8/CXCL8, ENA-78/CXCL5, groα/CXCL1, CTAP-III/CXCL7, GCP-2/CXCL6
CXCR3	IP-10/CXCL10, PF4/CXCL4, Mig/CXCL9
CXCR4 (fusin)	SDF-1/CXCL12
CXCR5	BCA-1/CXCL13
CXCR6	CXCL16
C-C chemokine receptors	
CCR1	MIP-1α/CCL3, RANTES/CCL5, MCP-3/CCL7, HCC-1/CCL14, HCC-2/CCL15, HCC-4/CCL16
CCR2	MCP-1/CCL2, MCP-3/CCL7, HCC-4/CCL16
CCR3	RANTES/CCL5, MCP-2/CCL8, MCP-3/CCL7, HCC-2/CCL15
CCR4	TARC/CCL17, CKLF1
CCR5	MIP-1α/CCL3, RANTES/CCL5, MCP-2/CCL8, HCC-1/CCL14
CCR6	MIP-3α/CCL20
CCR7	SLC/CCL21
C chemokine receptors	
XCR1	Lymphotactin/XCL1
C-X3-C chemokine receptors	
CX3CR1	Fractalkine/CX$_3$CL1
Other	
DARC	Duffy antigen, some CC and CXC chemokines

See text for abbreviations.

Among CXC chemokines, IL-8/CXCL8, ENA-78/CXCL5, groα/CXCL1, CTAP-III/CXCL7, granulocyte chemotactic protein 2 (GCP-2)/CXCL6, IP-10/CXCL10, Mig/CXCL9, PF4/CXCL4, SDF-1/CXCL12, as well as, recently, B cell-activating chemokine 1 (BCA-1)/CXCL13 and CXCL16 have been implicated in RA [1–5, 18–24].

Large amounts of IL-8/CXCL8 have been detected in the sera, synovial fluids (SF) and ST of RA patients [21, 25, 26]. Synovial macrophages constitutively express this chemokine, while proinflammatory cytokines, such as tumor necrosis factor-α (TNF-α) and IL-1, stimulate IL-8/CXCL8 production by fibroblasts [21, 27]. IL-8/

CXCL8 exerts proinflammatory and angiogenic effects [28]. Vascular endothelial cells (EC) express CXCR2, a receptor for the ELR+, angiogenic IL-8/CXCL8 [29].

ENA-78/CXCL5, similar to IL-8/CXCL8, is chemotactic for neutrophils and, being ELR+, stimulates angiogenesis [7, 17, 26]. Abundant ENA-78/CXCL5 has been detected in RA SF and ST [20]. Pro-inflammatory cytokines further augment the release of this chemokine [20]. Macrophages, fibroblasts and EC express ENA-78/CXCL5 [20].

groα/CXCL1, a neutrophil chemoattractant, has been detected in the SF and ST of RA patients [18, 27]. TNF-α or IL-1 stimulate the production of this chemokine [18, 27]. Synovial macrophages and fibroblasts express groα/CXCL1 [18].

The platelet-derived CTAP-III/CXCL7 has been detected in RA sera and ST [30]. This chemokine stimulates fibroblast proliferation of ST fibroblasts, proteoglycan synthesis and synovial fibrosis [30]. CTAP-III/CXCL7 also induces angiogenesis [1, 30].

GCP-2/CXCL6 expression is up-regulated on RA fibroblasts *via* Toll-like receptor 2 (TLR2) signaling [31]. There is abundant production of this chemokine in RA [31].

IP-10/CXCL10, Mig/CXCL9 and PF4/CXCL4 exert proinflammatory, but anti-angiogenic effects in RA [1, 17, 32–34]. These chemokines have been detected in the sera, SF and ST of RA patients [32–34]. RA fibroblasts and macrophages express these chemokines [32, 33]. The ELR-lacking IP-10/CXCL10, Mig/CXCL9 and PF4/CXCL4 suppress synovial neovascularization [17, 32, 34].

SDF-1/CXCL12 has unique properties in comparison to the CXC chemokines described above. First, while many of the other chemokines have common receptors, SDF-1/CXCL12 is a specific ligand for CXCR4. Moreover, this otherwise homeostatic chemokine has been implicated in T, B cell, and monocyte, rather than neutrophil, recruitment into the ST [22, 35]. Increased plasma levels of SDF-1/CXCL12 have been detected in RA patients in comparison to controls [36]. Finally, although this chemokine lacks the ELR motif, it is still angiogenic [37]. SDF-1/CXCL12 is also involved in cell adhesion. This chemokine is expressed by synovial EC and induces integrin-mediated adhesion of T cells [22], and binds to surface proteoglycans on EC [38]. Furthermore, direct cellular contact between T cells and fibroblasts induces SDF-1/CXCL12 production by the latter cells [39]. SDF/CXCL12 has been associated with osteoclastogenesis, bone resorption and thus increased radiographic progression in RA [40, 41].

The crucial role of B cells in the pathogenesis of RA has been acknowledged, and B cell targeting using rituximab has become a part of biological therapy in RA, as well as in lymphomas [12, 13, 42]. Some homeostatic chemokines involved in B cell migration and lymphoid tissue organization may also play a role in the pathogenesis of RA. B cell-activating chemokine-1 (BCA-1)/CXCL13 is the specific ligand for CXCR5 [9]. BCA-1/CXCL13 is a homeostatic chemokine involved in the migration of B cells and a subset of T cells, as well as in the formation of germinal centers

[1, 9, 10]. However, this chemokine is also expressed on follicular dendritic cells, as well as EC and fibroblasts in the RA ST [43]. Thus, BCA-1/CXCL13 has been implicated in inflammatory lymphoid tissue organization and aggregate formation in the RA ST [43].

CXCL16, the single specific ligand for CXCR6, may also be considered a homeostatic chemokine as it mediates lymphocyte recruitment to lymph nodes. Large amounts of CXCL16 were detected in RA SF and ST [23, 24]. Synovial macrophages and fibroblasts release this chemokine [23, 24]. CXCL16 recruits mononuclear cells to the RA ST [23]. CXCL16-mediated cell recruitment to the ST is dependent upon the MAP kinase pathway [23].

CC chemokines

CC chemokines contain adjacent CC residues and they stimulate monocyte chemotaxis. However, some CC chemokines also recruit lymphocytes [1, 2, 6, 8].

Among CC chemokines, monocyte chemoattractant protein-1 (MCP-1)/CCL2, macrophage inflammatory protein 1α (MIP-1α)/CCL3, MIP-3α/CCL20, 'regulated upon activation, normal T cell expressed and secreted' (RANTES)/CCL5, Epstein-Barr virus-induced gene 1 ligand chemokine (ELC)/CCL19, secondary lymphoid tissue chemokine (SLC)/CCL21 and, recently, chemokine-like factor 1 (CKLF1) have been implicated in inflammatory mechanisms underlying RA [1–4, 43, 44]. According to one recent study, MCP-2/CCL8, MCP-3/CCL7, hemofiltrate CC chemokine-1 (HCC-1)/CCL14, HCC-2/CCL15 and HCC-4/CCL16 may also be involved in RA [45].

MCP-1/CCL2 chemoattracts monocytes, T and natural killer (NK) cells [6, 46, 47]. Large amounts of this chemokine have been detected in RA sera, SF and ST [27, 46, 47]. Synovial macrophages and fibroblasts express MCP-1/CCL2 [27, 46, 47]. TNF-α, IL-1 and TLR2 ligands further stimulate the release of MCP-1/CCL2 by fibroblasts [27, 47].

MIP-1α/CCL3 chemoattracts monocytes, T, B and NK cells [2, 6]. This chemokine has been detected in SF and ST of RA patients [27, 48]. Synovial macrophages and fibroblasts produce MIP-1α/CCL3 [27, 48]. The release of MIP-1α/CCL3 is further augmented by proinflammatory cytokines [48].

MIP-3α/CCL20, the specific ligand for CCR6, is chemotactic for mononuclear cells, as well as immature dendritic cells [1, 49]. Abundant MIP-3α/CCL20 has been detected in RA SF and ST [49]. Synovial mononuclear cells and fibroblasts express this chemokine [49]. In a recent study, MIP-3α/CCL2 induced both osteoblast proliferation and osteoclast differentiation. An increased expression of this chemokine was detected in the subchondral bone tissue of RA patients. MIP-3α/CCL20 may act in concert with the RANKL system and may lead to uncoupling between new bone formation and bone resorption in RA [50].

RANTES/CCL5 exerts chemotactic activity towards monocytes, T and NK cells [6, 51]. RANTES/CCL5 has been detected in RA sera, SF and ST [32, 51]. Synovial T cells, macrophages and fibroblasts produce this chemokine [51, 52]. TNF-α and IL-1 augments RANTES/CCL5 production by fibroblasts [52]. A distinct polymorphism in the RANTES promoter gene has been associated with susceptibility to RA in Chinese patients [53].

ELC/CCL19 is homeostatic; however, it acts similarly to SDF-1/CXCL12 and BCA-1/CXCL13 in RA. ELC/CCL19 has been detected in RA ST [54].

SLC/CCL21 is another homeostatic chemokine. The production of this chemokine has been associated with the formation of lymphoid aggregates and germinal center-like structures in the RA ST [43].

Thymus and activation regulated chemokine (TARC)/CCL17 is a homeostatic chemokine also implicated in T cell recruitment into the ST. High levels of this chemokine have been detected in RA [55].

Pulmonary and activation regulated chemokine (PARC)/CCL18 is a chemokine with mostly homeostatic properties. High levels of this chemokine were detected in RA sera and SF. PARC/CCL18 is expressed in the RA ST, as well as in RA articular cartilage. Serum levels of PARC/CCL18 have been correlated with rheumatoid factor production [56, 57]. IL-4, IL-10, IL-13 and RA SF synergistically induce PARC/CCL18 production by monocytes [57].

In one study, MCP-2/CCL8, MCP-3/CCL7, HCC-1/CCL14, HCC-2/CCL15 and HCC-4/CCL16 were detected in the RA ST. Among these chemokines, HCC-2/CCL15 showed an increased expression in RA compared to osteoarthritic ST [45]. The function of these chemokines in RA needs to be confirmed by further studies.

CKLF1 is a unique cytokine chemotactic for various leukocytes. CKLF1 is a functional ligand of CCR4. Its expression is up-regulated on activated CD4$^+$ and CD8$^+$ T cells, but not on B cells in RA [44].

C and CX$_3$C chemokines

Members of these two chemokine subsets exert a special position of C residues [2, 58]. The C family contains two members, lymphotactin/XCL1 and single C motif 1β (SCM-1β)/XCL2. The CX$_3$C subset contains a single member, fractalkine/CX$_3$CL1 [2, 3, 58, 59].

Lymphotactin/XCL1 is primarily involved in T cell migration to inflammatory sites [1, 60]. This chemokine has been detected on CD8$^+$ and CD4$^+$/CD28$^-$ T cells in RA [60]. Lymphotactin/XCL1 augments T cell ingress into the RA joint [60].

Fractalkine/CX$_3$CL1 is chemotactic for mononuclear cells and also mediates cell adhesion [58, 61]. This chemokine has been detected in RA SF and ST [61, 62]. Synovial macrophages, fibroblasts, EC and dendritic cells produce this chemokine in RA [61, 62]. Fractalkine/CX$_3$CL1 also enhances the adhesion of CD4$^+$ T cells to fibroblasts [62]. Increased serum levels of fractalkine/CX_3CL1 have been associated with disease activity in rheumatoid vasculitis [63]. Fractalkine/CX_3CL1 has been implicated in angiogenesis and accelerated atherosclerosis associated with RA [64, 65]. A single nucleotide polymorphism (SNP) in the CX_3CR1 gene has been associated with reduced cardiovascular risk [64].

Chemokine receptors in RA

Chemokines are recognized by seven-transmembrane domain receptors expressed on the target cells [8]. There is a redundancy between CXC and CC chemokine receptors and their ligands (Tab. 1). Some receptors, such as CXCR2, CCR1 or CCR3 have numerous chemokine ligands, while CXCR4, CXCR5, CXCR6, CCR8 or CCR9 are specific receptors for one single ligand [1, 2, 8].

There may be a relationship between a certain chemokine receptor and the function of its ligand(s). For example, CXCR2, a receptor recognizing most ELR^+ CXC chemokines, plays a crucial role in inflammation and angiogenesis. In contrast, CXCR3 is a receptor for most ELR^-, angiostatic CXC chemokines [1, 2, 8, 17] (Tab. 1). Chemokine receptors have also been associated with various types of inflammation. For example, RA, which is considered a Th1-type disease, is associated with CXCR3 and CCR5, while asthma, a known Th2-type disease, is rather associated with CCR3, CCR4 and CCR8 [8, 61, 66, 67].

Among CXC chemokine receptors, CXCR1 and CXCR2 are receptors for the most important proinflammatory and pro-angiogenic CXC chemokines described above [1, 2] (Tab. 1). Both CXCR1 and CXCR2 are expressed on RA synovial macrophages and neutrophils, as well as articular chondrocytes [1, 2]. CXCR2, a receptor for most ELR^+ chemokines, is expressed by EC, and plays a role in chemokine-induced angiogenesis [29]. CXCR3 may be the most important receptor in leukocyte homing into the RA ST. CXCR3 is expressed in T cell rich areas of RA ST [61, 66, 68]. CXCR3 is also expressed on RA ST EC and dendritic cells [68]. CXCR4, the specific receptor for SDF-1/CXCL12, may play a role in the SDF-1/CXCL12-derived retention of lymphocytes within the RA ST [22]. CXCR5 is expressed by T cells, B cells, macrophages and EC in the RA ST [69]. CXCR6, the specific receptor for CXCL16, is expressed by one half of RA SF lymphocytes [23]. As described above, CXCR4, CXCR5 and CXCR6 bind their respective homeostatic chemokine ligands, SDF-1/CXCL12, BCA-1/CXCL13 and CXCL16. Thus, these CXC chemokine receptors may play an important role in lymphocyte recruitment under both homeostatic and inflammatory conditions [1, 5, 9, 22–24].

Among CC chemokine receptors, CCR1 is abundantly expressed in the RA ST [45, 70]. CCR5 show strong expression on RA ST T cells and fibroblasts [45, 61, 66]. CCR5, as well as CCR1, CCR2 and CCR3 are also expressed by articular chondrocytes [71]. CCR4 and CCR5 expressed on lymphocytes are crucial for leukocyte ingress into the RA joint [72]. There is an increasing body of evidence for the role of the truncated Δ32-CCR5 non-functional receptor allele in RA. This polymorphism of CCR5 may be protective against the development of RA [73]. CCR6, the single receptor for MIP-3α/CCL20 has been detected on RA ST leukocytes [49]. A putative chemokine receptor, CCR-like receptor 2 (CCRL2) has been identified on RA SF neutrophils and macrophages [74].

Regarding the C and CX_3C chemokine receptors, XCR1 is expressed on RA ST lymphocytes, macrophages and fibroblasts [1, 2]. CX3CR1 has been detected on macrophages and dendritic cells in the RA ST [59].

The receptor for the Duffy blood group antigen, DARC, also recognizes some CXC and CC chemokines (Tab. 1) [75]. DARC is expressed on RA ST EC [75].

Inflammatory and homeostatic chemokines in RA

As several functions of the chemokines described above may overlap in RA, it is highly debatable whether such functional classification is really justified. However, some of these mediators are primarily involved in the effector functions underlying inflammation, while others may also play a role in lymphoid tissue reorganization in the synovium (for reviews see [1–4, 43, 44]).

Most CXC and CC chemokines, as well as all C and CX_3C chemokines implicated in the pathogenesis of RA are definitely inflammatory chemokines [9]. As described above, the CXC chemokines IL-8/CXCL8, ENA-78/CXCL5, groα/CXCL1, CTAP-III/CXCL7, IP-10/CXCL10, Mig/CXCL9, PF4/CXCL4, GCP-2/CXCL6, SDF-1/CXCL12, BCA-1/CXCL13 and CXCL16, the CC chemokines MCP-1/CCL2, MIP-1α/CCL3, MIP-3α/CCL20, RANTES/CCL5, ELC/CCL19 and SLC/CCL21, as well as lymphotactin/XCL1 and fractalkine/CX_3CL1 have been implicated in leukocyte recruitment underlying inflammatory synovitis (for reviews see [1–4, 43, 44]) (Tab. 1). Accordingly, CXCR1-6, CCR1-6, as well as XCR1 and CX_3CR1, receptors that mostly recognize inflammatory chemokine ligands, have also been implicated in RA (for reviews see [1–4]). In addition, as discussed above, numerous inflammatory CXC chemokines are also angiogenic in RA (for reviews see [1–4]).

Homeostatic chemokines are constitutively produced in discrete microenvironments of lymphoid or non-lymphoid tissues. These chemokines are involved in the physiological traffic of cells into tissues during the immune surveillance [9, 10]. However, as discussed above, some data suggest that, among these homeostatic chemokines, SDF-1/CXCL12, BCA-1/CXCL13, CXCL16, TARC/CCL17, PARC/CCL18, ELC/CCL19, SLC/CCL21 and maybe others are also involved in RA-associated inflammatory cell recruitment [1, 19, 22-24, 37, 43, 54]. The synovium, in many ways, is similar to the skin- and mucosa-associated lymphoid tissues, which may explain the involvement of classical homeostatic chemokines in synovial inflammation [9, 10, 43].

Recent advances in the regulation of chemokine production

As described above, chemokine secretion and chemokine receptor expression may be regulated by other inflammatory mediators, such as pro-inflammatory cytokines [1–4, 11].

Autoimmunity and immune complexes may also trigger chemokine production. In a recent study, immune complexes containing antibodies to type II collagen from arthritis patients induced IL-8/CXCR8 release by monocytes *via* an Fcγ receptor IIA-dependent mechanism [76].

Increased chemokine production may be associated with the production of autoantibodies in RA. Patients with RA were identified from a population of blood donors. Analysis of serum samples antedated the onset of RA by a median of 3 years. Serum MCP-1/CCL2 levels were increased in IgM rheumatoid factor (RF)- and anti-cyclic citrullinated peptide (anti-CCP)-positive subjects, as well as in patients later developing RA. Thus, increased chemokine secretion, in association with IgM RF and anti-CCP production, may precede the onset of RA by years [77].

Intercellular adhesion itself may also promote chemokine release from cells. For example, cellular contact between T cells and fibroblasts induces SDF-1/CXCL12 production by the fibroblasts [39].

Targeting of chemokines and chemokine receptors

Chemokines and chemokine receptors may be targeted by nonspecific strategies and by specific approaches. Until now, there have been few human RA trials, therefore most available data have been obtained in animal models of arthritis (for reviews see [1–4, 11]) (Tab. 2).

Nonspecific blockade of chemokine production and chemokine receptor expression

Corticosteroids, such as dexamethasone, effectively suppressed IL-8/CXCL8 and MCP-1/CCL2 production in RA [78]. Among non-steroidal anti-inflammatory drugs (NSAID), diclofenac and meloxicam attenuated IL-8/CXCL8 production in rat antigen-induced arthritis (AgIA) [79].

Among disease-modifying antirheumatic drugs (DMARD), sulfasalazine inhibited the release of IL-8/CXCL8, groα/CXCL1 and MCP-1/CCL2 by cultured RA ST explants [80]. A combination treatment of RA patients with methotrexate and leflunomide decreased MCP-1/CCL2 expression in the ST. Suppression of chemokine release was associated with some clinical improvement [81]. Methotrexate also decreased CCR2 expression on peripheral blood monocytes obtained

Table 2 - Chemokine and chemokine receptor targeting strategies in RA. *

Compound	Chemokine/chemokine receptor	Animal (A)/ Human (H) study**	Reference(s)
Nonspecific blockade			
Dexamethasone	IL-8/CXCL8, MCP-1/CCL2	H	[78]
NSAIDs	IL-8/CXCL8	A	[79]
Sulfasalazine	IL-8/CXCL8, groα/CXCL1, MCP-1/CCL2	H	[80]
Methotrexate	CCR2	H	[82]
Methotrexate + leflunomide	MCP-1/CCL2	H	[81]
Infliximab	IL-8/CXCL8, RANTES/CCL5, MCP-1/CCL2, groα/CXCL1, CXCL16	H	[83–86, 90]
Etanercept	CXCR3	H	[88]
Antioxidants	IL-8/CXCL8, MCP-1/CCL2	H	[91]
Simvastatin	IL-8/CXCL8	H	[92]
Triptolide	MCP-1/CCL2, RANTES/CCL5	A	[93]
Epigallocatechin-3-gallate	ENA-78/CXCL5, groα/CXCL1, RANTES/CCL5	H	[94]
COX/LOX inhibitor (ML3000)	Míg/CXCL9, IP-10/CXCL10, I-TAC/CXCL11	H	[95]
PPARγ agonists	MCP-1/CCL2	H	[96]
Specific blockade			
Anti-ENA-78/CXCL5	ENA-78/CXCL5	A	[98]
Anti-groα /CXCL1	groα /CXCL1	A	[99]
Anti-IP-10/CXCL10	IP-10/CXCL10	A	[100]
PF4/CXCL4 peptide	PF4/CXCL4	A	[19]
Anti-CXCL16	CXCL16	A	[24]
Anti-IL-8/CXCL8	IL-8/CXCL8	A	[97]
Anti-MIP-1α/CCL3	MIP-1α/CCL3	A	[99]
Anti-MCP-1/CCL2	MCP-1/CCL2	A	[102, 104, 105]
Anti-RANTES/CCL5	RANTES/CCL5	A	[106]
KE-298	MCP-1/CCL2 + RANTES/CCL5	A	[107]
ABN912	MCP-10/CCL2	H	[108]

Table 2 (continued)

Compound	Chemokine/chemokine receptor	Animal (A)/ Human (H) study**	Reference(s)
Antibody	Fractalkine/CX3CL1	A	[109]
Combined blockade	Various chemokines together	A	[111–113]
CXCR2 antagonist	CXCR2	A	[114]
DF2162	CXCR1 + CXCR2	A	[115]
TAK-779	CXCR3 + CCR5	A	[117]
AMD3100	CXCR4	A	[118]
FC131, other T140 analogues	CXCR4	A	[118, 119]
J-113863	CCR1	A	[125]
CP-481,715	CCR1	H	[121, 122, 132]
MetRANTES	CCR1 + CCR5	A	[126, 127]
CCR5 antagonist	CCR5	A	[128]
Anti-CCR2 antibodies	CCR2	A	[123, 129]
MLN1202	CCR2	H	[133]
Maraviroc	CCR5	H	[134]

See text for abbreviations.
**Human studies include both* in vitro *studies with isolated cells or histological samples, as well as* in vivo *studies.*

from RA patients. Decrease in cell surface CCR2 density was associated with lower disease activity [82].

Among anti-TNF-α agents, infliximab inhibited the expression of IL-8/CXCL8, RANTES/CCL5 and MCP-1/CCL2 in RA sera and ST [83, 84]. Infliximab also suppressed serum levels of groα/CXCL1 and CXCL16 in RA [85, 86]. In another study, infliximab reduced the expression of CCR3 and CCR5 on RA T cells [87]. Treatment of RA patients with infliximab or etanercept resulted in the clearance of CXCR3$^+$ T cells from the ST [88]. Anti-TNF therapy also attenuated CXCL16 expression on synovial macrophages [89]. Chemokine inhibition may also have relevance for safety of anti-TNF therapy as infliximab reduced the secretion of IL-8/CXCL8, MIP-1α/CCL3 and MCP-1/CCL2 in response to mycobacteria. These results suggest that the increased incidence of tuberculosis in infliximab-treated RA patients may be related, in part, to the inhibition of TNF-dependent chemokine gradients and abnormal leukocyte migration [90].

Some antioxidants, such as *N*-acetyl-L-cysteine and 2-oxothiazolidine-4-carboxylate, inhibited the expression of IL-8/CXCL8 and MCP-1/CCL2 mRNA by cytokine-treated isolated human synovial fibroblasts [91]. Statins may exert anti-inflammatory effects. In a recent study, simvastatin inhibited the release of IL-8/CXCL8 by TNF-α-stimulated synovial fibroblasts obtained from RA patients [92]. Triptolide, a diterpenoid triepoxide with potent anti-inflammatory effects, inhibited MCP-1/CCL2, MIP-1α/CCL3 and RANTES/CCL5 protein production and mRNA expression in the rat adjuvant-induced arthritis (AIA) model [93]. Epigallocatechin-3-gallate (EGCG), a potent anti-inflammatory compound derived from green tea, suppressed IL-1-induced ENA-78/CXCL5, groα/CXCL1 and RANTES/CCL5 production by RA ST fibroblasts [94]. A dual cyclooxygenase-lipoxygenase (COX/LOX) inhibitor, ML3000, inhibited Mig/CXCL9, IP-10/CXCL10 and I-TAC/CXCL11 expression on RA synovial fibroblasts [95]. Activation of peroxisome proliferator-activated receptor γ (PPARγ) suppresses MCP-1/CCL2 expression in monocytes [96]. Thus, PPARγ agonists, such as glitazones, may inhibit chemokine production.

Specific targeting of chemokines

Regarding CXC chemokines, neutralizing antibodies to IL-8/CXCL8 prevented arthritis in rabbits [97]. A neutralizing polyclonal anti-ENA-78/CXCL5 antibody was administered intravenously to rats using the AIA model. The antibody administered preventatively to the animals attenuated the severity of the disease; however, this antibody was ineffective when administered therapeutically [98]. The preventative administration of an anti-groα/CXCL1 antibody delayed the onset and severity of murine collagen-induced arthritis (CIA) [99]. Adoptive transfer of self-specific anti-IP-10/CXCL10 or rabbit anti-rat IP-10/CXCL10 antibodies resulted in the suppression of rat AIA [100]. A bioactive synthetic peptide derived from PF4/CXCL4 inhibited the development of murine CIA [19]. An anti-CXCL16 monoclonal antibody reduced the clinical arthritis score and suppressed joint destruction in murine CIA [24].

There has been one human trial using an anti-IL-8/CXCL8 antibody in RA, but results of this trial were not published and the further development of this antibody was terminated (for review see [1]). Yet, 10F8, a monoclonal antibody to IL-8 reduced clinical activity in palmoplantar pustulosis [101], a chronic inflammatory disease that may be associated with various types of arthritis.

Passive immunization of mice with anti-MIP-1α/CCL3 decreased the severity of murine CIA [99]. MIP-1α/CCL3 gene-deficient mice exhibit a milder course of CIA [102]. A monoclonal antibody to MCP-1/CCL2 reduced ankle swelling and decreased the number of synovial macrophages in rat CIA [103]. An anti-MCP-1/CCL2 antibody also prevented the recruitment of ^{111}In-labeled T cells into the synovium in the rat model of streptococcal cell wall (SCW) antigen-induced arthritis

[104]. A peptide inhibitor of MCP-1/CCL2 suppressed the development of arthritis in MRL-lpr mice [105]. An antibody to RANTES/CCL5 attenuated murine CIA [106]. KE-298 inhibited both MCP-1/CCL2 and RANTES/CCL5 production, as well as the severity of rat AIA [107].

Regarding human trials targeting CC chemokines, a randomized, controlled trial has recently been conducted using ABN912, a monoclonal antibody to MCP-1/CCL2. In this study, 33 patients received the active compound, and 12 received placebo. Serial arthroscopic biopsies were performed. ABN912 treatment was well tolerated, but there was no detectable clinical benefit or significant change in ST biomarkers. In addition, there was a dose-related, 2000-fold increase in serum MCP-1/CCL2 levels [108].

Regarding other chemokines, an antibody to fractalkine/CX3CL1 decreased clinical severity and joint destruction in murine CIA [105].

Viral gene transfer may also be used to target chemokines. A 35-kDa soluble protein (35k) derived from vaccinia virus inactivates numerous CC chemokines. A recombinant adenovirus containing 35k reduced migration of CCR5-transfected cells in response to RANTES/CCL5. This vector also suppressed CCR5-dependent chemotaxis of murine macrophages [110].

Combined chemokine blockade has been used in some recent studies. A combination of MCP-1/CCL2 and groα/CXCL1 inhibition resulted in a greater extent of arthritis suppression than MCP-1/CCL2 blockade alone in murine AIA [111]. Rat AIA was suppressed by DNA vaccination using a combination of chemokine DNA vaccines to MCP-1/CCL2, MIP-1α/CCL3 and RANTES/CCL5 [112]. In the rabbit endotoxin-induced arthritis model, the combination of anti-IL-8/CXCL8 and anti-groα/CXCL1 antibodies resulted in a more pronounced inhibition of knee arthritis than did any of the two antibodies alone [113]. Certainly, an increased toxicity using combined strategies may be an important issue during future human trials [1].

Selective inhibition of chemokine receptors

A nonpeptide oral antagonist of the CXCR2 receptor inhibited IL-8/CXCL-induced arthritis in rabbits [114]. DF2162, an allosteric CXCR1/CXCR2 inhibitor diminished murine and rat arthritis [115]. CXCR2 gene-deficient mice exerted attenuated severity of Lyme arthritis in comparison to wild-type animals [116]. TAK-779, a common inhibitor of CXCR3 and CCR5 suppressed the binding of these receptors to their respective chemokine ligands, as well as chemokine-induced integrin activation [117]. AMD3100, a CXCR4 antagonist, inhibited CIA in IFN-γ-deficient mice [118]. Analogs of the 14-mer peptide T140, acting as CXCR4 inhibitors, ameliorated murine CIA [119]. Other low molecular weight CXCR4 antagonists, such as FC131, have been developed by exploratory structural tuning of cyclic polypeptide scaffolds [119].

Numerous CCR1 and CCR2 antagonists have been developed and preclinical studies with chemokine receptor-deficient mice have been initiated during the last decade [120–125]. J-113863, a small molecule CCR1 antagonist improved paw inflammation and joint destruction in murine CIA [125]. Preventatively or therapeutically administered Met-RANTES, a CCR1/CCR5 antagonist, inhibited murine CIA and rat AIA. Met-RANTES also suppressed CCR1 and CCR5 expression in the joint [126, 127]. A nonpeptide CCR5 antagonist preventatively inhibited mouse CIA [128]. Some anti-chemokine receptor effects may be dose-dependent. For example, while low doses of the MC-21 anti-CCR2 monoclonal antibody markedly improved murine CIA, high doses of this antibody rather had pro-inflammatory effects [129].

Studies with gene-deficient *versus* antagonist-treated animals resulted in somewhat conflicting results. For example, while a non-peptide CCR1 antagonist improved murine CIA, CCR1-deficient mice produced more TNF-α than controls [125]. In one study, CCR5$^{-/-}$ mice developed CIA to the same extent as wild-type animals [130]. However, in another study, CCR5 gene-deficient animals exerted a significant reduction in the incidence of CIA, which was associated with an increased IL-10 production by spleen cells [131]. Similar controversy was observed when comparing results obtained using anti-CCR2 antibodies or CCR2 gene-deficient animals. Anti-CCR2 antibody administered during the initiation of murine CIA markedly improved the clinical symptoms, while blockade during the later stages of the disease aggravated arthritis [123]. CCR2$^{-/-}$ mice developed more severe CIA than did wild-type controls [123, 124, 130]. These data suggest that data obtained from chemokine receptor antagonist studies in comparison to those obtained from studies using gene-deficient animals may not be comparable. Yet, CCR blockade using antibodies or synthetic inhibitors may be promising for future therapies.

Regarding the limited number of human RA trials with chemokine receptor antagonists, a CCR1 antagonist has been tried in a 2-week Phase Ib study. This inhibitor decreased the number of ST macrophages [122]. One-third of the patients also fulfilled the ACR20% criteria for clinical improvement [122]. CP-481,715, a selective CCR1 antagonist inhibited monocyte chemotactic activity present in human RA SF samples [121]. This compound has been evaluated in Phase I for pharmacokinetics and safety [132].

Some CCR2 inhibitors have also entered clinical trials [123]. In a recent Phase IIa clinical trial, MLN1202, a human CCR2 blocking antibody was administered to RA patients. Treatment with this blocking antibody reduced the levels of free CCR2 on CD14$^+$ monocytes by 57–94% demonstrating the biological activity of this agent. However, no clinical improvement was observed, suggesting that CCR2 blockade itself may not be sufficient to control synovitis in RA [133].

Among CCR5 inhibitors, maraviroc has been introduced to Phase II-III trials in HIV infection and AIDS, as well as to Phase II trial in RA [134].

Conclusions

In this chapter, we have discussed the putative role of chemokines and their receptors in RA. We also presented some examples for recent chemokine and chemokine receptor targeting strategies. There is a structural and a functional classification of chemokines. The former includes four groups: CXC, CC, C and CX_3C chemokines. Chemokines may also be distinguished as being inflammatory or homeostatic; however, some chemokines exert overlapping functions. Anti-chemokine and anti-chemokine receptor targeting may be therapeutically used in the future biological therapy of arthritis. In addition to the clear clinical benefit, we can learn a lot from these trials about the actions of the targeted chemokines and their receptors. Until now, most data in this field have been obtained from animal models of arthritis as only very few human RA trials have been completed. However, it is very likely that numerous specific chemokine and chemokine receptor antagonists will be developed, administered to RA patients and some caveats discussed above will be clarified in the near future.

Acknowledgements

This work was supported by NIH grants AR-048267 and AI-40987 (A.E.K.), the William D. Robinson and Frederick Huetwell Endowed Professorship (A.E.K.), Funds from the Veterans' Administration (A.E.K.); and grant No T048541 from the National Scientific Research Fund (OTKA) (Z.S.).

References

1   Koch AE (2005) Chemokines and their receptors in rheumatoid arthritis. *Arthritis Rheum* 52: 710–721

2   Szekanecz Z, Kim J, Koch AE (2002) Chemokines and chemokine receptors in rheumatoid arthritis. *Semin Immunol* 399: 1–7

3   Szekanecz Z, Koch AE (2001) Chemokines and angiogenesis. *Curr Opin Rheumatol* 13: 202–208

4   Szekanecz Z, Szücs G, Szántó S, Koch AE (2006) Chemokines in rheumatic diseases. *Curr Drug Targets* 7: 91–102

5   Vergunst CE, Tak PP (2005) Chemokines: Their role in rheumatoid arthritis. *Curr Rheumatol Rep* 7: 382–388

6   Taub DD (1996) C-C chemokines – an overview. In: AE Koch, RM Strieter (eds): *Chemokines in Disease*. RG Landes Company, Austin, pp 27–54

7   Walz A, Kunkel SL, Strieter RM (1996) C-X-C chemokines – an overview. In: AE Koch, RM Strieter (eds): *Chemokines in Disease*. RG Landes Company, Austin, pp 1–25

8   Zlotnik A, Yoshie O (2000) Chemokines: A new classification system and their role in immunity. *Immunity* 12: 121–127

9   Moser B, Loetscher P (2001) Lymphocyte traffic control by chemokines. *Nat Immunol* 2: 123–128

10   Kunkel EJ, Butcher EC (2002) Chemokines and the tissue-specific migration of lymphocytes. *Immunity* 16: 1–4

11   Tak PP (2006) Chemokine inhibition in inflammatory arthritis. *Best Pract Res Clin Rheumatol* 20: 929–939

12   Silverman GJ, Carson DA (2003) Roles of B cells in rheumatoid arthritis. *Arthritis Res Ther* 5 Suppl 4: 1–6

13   De Vita S, Zaja F, Sacco S, De Candia A, Fanin R, Ferraccioli G. (2002) Efficacy of selective B cell blockade in the treatment of rheumatoid arthritis. *Arthritis Rheum* 46: 2029–2033

14   Harris ED (1990) Rheumatoid arthritis: Pathophysiology and implications for therapy. *N Engl J Med* 332: 1277–1287

15   Oppenheim JJ, Zachariae COC, Mukaida N, Matsushima K (1991) Properties of the novel proinflammatory supergene "intercrine" cytokine family. *Annu Rev Immunol* 9: 617–648

16   Bacon K, Baggiolini M, Broxmeyer H, Horuk R, Lindley I, Mantovani A, Maysushima K, Murphy P, Nomiyama H, Oppenheim J et al (2002) Chemokine/chemokine receptor nomenclature. *J Interferon Cytokine Res* 22: 1067–1068

17   Strieter RM, Polverini PJ, Kunkel SL, Arenberg DA, Burdick MD, Kasper J, Dzuiba J, Van Damme J, Walz A, Marriott D et al (1995) The functional role of the ELR motif in CXC chemokine-mediated angiogenesis. *J Biol Chem* 270: 27348–27357

18   Koch AE, Kunkel SL, Shah MR, Hosaka S, Halloran MM, Haines GK, Burdick MD, Pope RM, Strieter RM (1995) Growth related gene product alpha: A chemotactic cytokine for neutrophils in rheumatoid arthritis. *J Immunol* 155: 3660–3666

19   Wooley PH, Schaefer C, Whalen JD. Dutcher JA, Counts JF (1997) A peptide sequence from platelet factor 4 (CT-112) is effective in the treatment of type II collagen induced arthritis in mice. *J Rheumatol* 24: 890–898

20   Koch AE, Kunkel SL, Harlow LA, Johnson B, Evanoff HL, Haines GK, Burdick MD, Pope RM, Strieter RM (1994) Epithelial neutrophil activating peptide-78: A novel chemotactic cytokine for neutrophils in arthritis. *J Clin Invest* 94: 1012–1018

21   Koch AE, Kunkel SL, Burrows JC, Evanoff HL, Haines GK, Pope RM, Strieter RM (1991) Synovial tissue macrophage as a source of the chemotactic cytokine IL-8. *J Immunol* 147: 2187–2195

22   Nanki T, Hayashida K, El-Gabalawy HS, Suson S, Shi K, Girschick HJ, Yavuz S, Lipsky PE (2000) Stromal cell-derived factor-1-CXC chemokine receptor 4 interactions play a central role in CD4$^+$ T-cell accumulation in rheumatoid arthritis synovium. *J Immunol* 165: 6590–6598

23   Ruth JH, Haas CS, Park CC, Amin MA, Martinez RJ, Haines GK, Shahrara S, Campbell PL, Koch AE (2006) CXCL16-mediated cell recruitment to rheumatoid arthritis syn-

ovial tissue and murine lymph nodes is dependent upon the MAPK pathway. *Arthritis Rheum* 54: 765–778

24 Nanki T, Shimaoka T, Hayashida K Taniguchi K, Yonehara S, Miyasaka N (2005) Pathogenic role of CXCL16-CXCR6 pathway in rheumatoid arthritis. *Arthritis Rheum* 52: 3004–3014

25 Deleuran B, Lemche P, Kristensen M, Chu CQ, Field M, Jensen J, Matsushima K, Stengaard-Pedersen K (1994) Localisation of interleukin 8 in the synovial membrane, cartilage–pannus junction and chondrocytes in rheumatoid arthritis. *Scand J Rheumatol* 23: 2–7

26 Koch AE, Volin MV, Woods JM, Kunkel SL, Connors MA, Harlow LA, Woodruff DC, Burdick MD, Strieter RM (2001) Regulation of angiogenesis by the C-X-C chemokines interleukin-8 and epithelial neutrophil activating peptide-78 in the rheumatoid joint. *Arthritis Rheum* 44: 31–40

27 Hosaka S, Akahoshi T, Wada C, Kondo H (1994) Expression of the chemokine super-family in rheumatoid arthritis. *Clin Exp Immunol* 97: 451–457

28 Koch AE, Polverini PJ, Kunkel SL, Harlow LA, DiPietro LA, Elner VM, Elner SG, Strieter RM (1992) Interleukin-8 as a macrophage-derived mediator of angiogenesis. *Science* 258: 1798–1801

29 Salcedo R, Ponce ML, Young HA, Wasserman K, Ward JM, Kleinman HK, Oppenheim JJ, Murphy WJ (2000) Human endothelial cells express CCR2 and respond to MCP-1: Direct role of MCP-1 in angiogenesis and tumor progression. *Blood* 96: 34–40

30 Castor CW, Smith EM, Hossler PA, Bignall MC, Aaron BP (1992) Detection of connective tissue activating peptide-III isoforms in synovium from osteoarthritis and rheumatoid arthritis patients: Patterns of interaction with other synovial cytokines in cell culture. *Arthritis Rheum* 35: 783–793

31 Pierer M, Rethage J, Seibl R, Lauener R, Brentano F, Wagner U, Hantzschel H, Michel BA, Gay RE, Gay S, Kyburz D (2004) Chemokine secretion of rheumatoid arthritis synovial fibroblasts stimulated by Toll-like receptor 2 ligands. *J Immunol* 172: 1256–1265

32 Hanaoka R, Kasama T, Muramatsu M Yajima N, Shiozawa F, Miwa Y, Negishi M, Ide H, Miyaoka H, Uchida H et al (2003) A novel mechanism for the regulation of IFN-γ inducible protein-10 expression in rheumatoid arthritis. *Arthritis Res Ther* 5: R74–R81

33 Patel DD, Zachariah JP, Whichard LP (2001) CXCR3 and CCR5 ligands in the rheumatoid arthritis synovium. *Clin Immunol* 98: 39–45

34 Blades MC, Ingegnoli F, Wheller SK, Manzo A, Wahid S, Panayi GS, Perretti M, Pitzalis C (2002) Stromal cell-derived factor 1 (CXCL12) induces monocyte migration into human synovium transplanted onto SCID mice. *Arthritis Rheum* 46: 824–836

35 Bradfield PF, Amft N, Vernon-Wilson E Exley AE, Parsonage G, Rainger GE, Nash GB, Thomas AM, Simmons DL, Salmon M et al (2003) Rheumatoid fibroblast-like synoviocytes overexpress the chemokine stromal cell-derived factor 1 (CXCL12) which

supports distinct patterns and rates of CD4$^+$ and CD8$^+$ T cell migration within synovial tissue. *Arthritis Rheum* 48: 2472–2482

36   Hansen IB, Ellingsen T, Hornung N, Poulsen JH, Lottenburger T, Stengaard-Pedersen K (2006) Plasma level of CXC-chemokine CXCL12 is increased in rheumatoid arthritis and is independent of disease activity and methotrexate treatment. *J Rheumatol* 33: 1754–1759

37   Pablos JL, Santiago B, Galindo M Torres C, Brehmer MT, Blanco FJ, Garcia-Lazaro FJ (2003) Synoviocyte-derived CXCL12 is displayed on endothelium and induces angiogenesis in rheumatoid arthritis. *J Immunol* 170: 2147–2152

38   Santiago B, Baleux F, Palao G, Gutierrez-Canas I, Ramirez JC, Arenzana-Seisdedos F, Pablos JL (2006) CXCL12 is displayed by rheumatoid endothelial cells through its basic amino-terminal motif on heparan sulfate proteoglycans. *Arthritis Res Ther* 8: R43

39   Burger D (2000) Cell contact interactions in rheumatology. The Kennedy Institute for Rheumatology, London, UK, 1–2 June 2000. *Arthritis Res* 2: 472–476

40   Joven B, Gonzalez N, Aguilar F, Santiago B, Galindo M, Alcami J (2005) Association between stromal cell derived factor 1 chemokine gene variant and radiographic progression of rheumatoid arthritis. *Arthritis Rheum* 52: 354–356

41   Gronthos S, Zannettino AC (2007) The role of the chemokine CXCL12 in osteoclastogenesis. *Trends Endocrinol Metab* 18: 108–113

42   Edwards JCW, Cambridge G, Abrahams VM (1999) Do self-perpetuating B lymphocytes drive human autoimmune disease? *Immunology* 97: 188–196

43   Manzo A, Paoletti S, Carulli M, Blades MC, Barone F, Yanni G, Fitzgerald O, Bresnihan B, Caporali R, Montecucco C (2005) Systematic microanatomical analysis of CXCL13 and CCL21 *in situ* production and progressive lymphoid organization in rheumatoid synovitis. *Eur J Immunol* 35: 1347–1359

44   Li T, Zhong J, Chen Y, Qiu X, Zhang T, Ma D, Han W (2006) Expression of chemokine-like factor 1 is upregulated during T lymphocyte activation. *Life Sci* 79: 519–524

45   Haringman JJ, Smeets TJ, Reinders-Blankert P, Tak PP (2006) Chemokine and chemokine receptor expression in paired peripheral blood mononuclear cells and synovial tissue of patients with rheumatoid arthritis, osteoarthritis and reactive arthritis. *Ann Rheum Dis* 65: 294–300

46   Koch AE, Kunkel SL, Harlow LA Johnson B, Evanoff HL, Haines GK, Burdick MD, Pope RM, Strieter RM (1992) Enhanced production of monocyte chemoattractant protein-1 in rheumatoid arthritis. *J Clin Invest* 90: 772–779

47   Villiger PM, Terkeltaub R, Lotz M (1992) Production of monocyte chemoattractant protein-1 by inflamed synovial tissue and cultured synoviocytes. *J Immunol* 149: 722–727

48   Koch AE, Kunkel SL, Harlow LA Mazarakis DD, Haines GK, Burdick MD, Pope RM, Strieter RM (1994) Macrophage inflammatory protein-1 alpha. A novel chemotactic cytokine for macrophages in rheumatoid arthritis. *J Clin Invest* 93: 921–928

49   Matsui T, Akahoshi T, Namai R, Hashimoto A, Kurihara Y, Rana M, Nishimura A, Endo H, Kitasato H, Kawai S et al (2001) Selective recruitment of CCR6-expressing

cells by increased production of MIP-3 alpha in rheumatoid arthritis. *Clin Exp Immunol* 125: 155–161

50   Lisignoli G, Piacentini A, Cristino S, Grassi F, Cavallo C, Cattini L, Tonnarelli B, Manferdini C, Facchini A (2007) CCL20 chemokine induces both osteoblast proliferation and osteoclast differentiation: Increased levels of CCL20 are expressed in subchondral bone tissue of rheumatoid arthritis patients. *J Cell Physiol* 210: 798–806

51   Volin MV, Shah MR, Tokuhira M, Haines GK, Woods JM, Koch AE (1998) RANTES expression and contribution to monocyte chemotaxis in arthritis. *Clin Immunol* Immunopathol 89: 44–53

52   Rathanaswami P, Hachicha M, Sadick M, Scahll TJ, McColl SR (1993) Expression of the cytokine RANTES in human rheumatoid synovial fibroblasts. Differential regulation of RANTES and interleukin-8 genes by inflammatory cytokines. *J Biol Chem* 268: 5834–5839

53   Wang CR, Guo HR, Liu MF (2005) RANTES promoter polymorphism as a genetic risk factor for rheumatoid arthritis in the Chinese. *Clin Exp Rheumatol* 23: 379–384

54   Buckley CD (2003) Why do leucocytes accumulate within chronically inflamed joints? *Rheumatology (Oxford)* 42: 1433–1444

55   Okamoto H, Koizumi K, Yamanaka H, Saito T, Kamatani N (2003) A role for TARC/CCL17, a CC chemokine, in systemic lupus erythematosus. *J Rheumatol* 30: 2369–2373

56   Momohara S, Okamoto H, Iwamoto T, Mizumura T, Ikari K, Kawaguchi Y, Takeuchi M, Kamatani N, Tomatsu T (2007) High CCL18/PARC expression in articular cartilage and synovial tissue of patients with rheumatoid arthritis. *J Rheumatol* 34: 266–271

57   Van Lieshout AW, van der Voort R, le Blanc LM, Roelofs MF, Schreurs BW, van Riel PL, Adema GJ, Radstake TR (2006) Novel insights in the regulation of CCL18 secretion by monocytes and dendritic cells *via* cytokines, toll-like receptors and rheumatoid synovial fluid. *BMC Immunol* 19: 7–23

58   Bazan JF, Bacon KB, Hardiman G, Wang W, Soo K, Rossi D, Greaves DR, Zlotnik A, Schall TJ (1997) A new class of membrane bound chemokine with a X3C motif. *Nature* 385: 640–644

59   Ruth JH, Volin MV, Haines GK III, Woodruff DC, Katschke KJ Jr, Woods JM, Park CC, Morel JC, Koch AE (2001) Fractalkine, a novel chemokine in rheumatoid arthritis and rat adjuvant-induced arthritis. *Arthritis Rheum* 44: 1568–1581

60   Blaschke S, Middel P, Dorner BG Blaschke V, Hummel KM, Kroczek RA, Reich K, Benoehr P, Koziolek M, Muller GA (2003) Expression of activation-induced, T cell-derived, and chemokine related cytokine/lymphotactin and its functional role in rheumatoid arthritis. *Arthritis Rheum* 48: 1858–1872

61   Loetscher P, Uguccioni M, Bordoli L, Baggiolini M, Moser B, Chizzolini C, Dayer JM (1998) CCR5 is characteristic of Th1 lymphocytes. *Nature* 391: 344–345

62   Sawai H, Park YW, Robertson J, Iwai T, Goronzy JJ, Weyand CM (2005) T cell costimulation by fractalkine-expressing synoviocytes in rheumatoid arthritis. *Arthritis Rheum* 52: 1392–1401

63   Matsunawa M, Isozaki T, Odai T, Yjima N, Takeuchi HT, Negishi M, Ide H, Adachi M, Kasama T (2006) Increased serum levels of soluble fractalkine (CX3CL1) correlate with disease activity in rheumatoid vasculitis. *Arthritis Rheum* 54: 3408–3416

64   McDermott DH, Fong AM, Yang Q Sechler JM, Cupples LA, Merrell MN, Wilson PW, D'Agostino RB, O'Donnell CJ, Patel DD et al (2003) Chemokine receptor mutant CX3CR1-M280 has impaired adhesive function and correlates with protection from cardiovascular disease in humans. *J Clin Invest* 111: 1241–1250

65   Volin MV, Woods JM, Amin MA, Connors MA, Harlow LA, Koch AE (2001) Fractalkine: A novel angiogenic chemokine in rheumatoid arthritis. *Am J Pathol* 159: 1521–1530

66   Qin S, Rottman JB, Myers P, Kassam N, Weinblatt M, Loetscher M, Koch AE, Moser B, Mackay CR (1998) The chemokine receptors CXCR3 and CCR5 mark subsets of T cells with a homing predilection for certain inflammatory sites. *J Clin Invest* 101: 746–750

67   Norii M, Yamamura M, Iwahashi M, Ueno A, Yamana J, Makino H (2006) Selective recruitment of CXCR3$^+$ and CCR5$^+$ T cells into synovial tissue in patients with rheumatoid arthritis. *Acta Med Okayama* 60: 149–157

68   Garcia-Lopez MA, Sanchez-Madrid F, Rodriguez-Frade JM (2001) CXCR3 chemokine receptor distribution in normal and inflamed tissues. *Lab Invest* 81: 409–418

69   Schmutz C, Hulme A, Burman A Salmon M, Ashton B, Buckley C, Middleton J (2005) Chemokine receptors in the rheumatoid synovium: Upregulation of CXCR5. *Arthritis Res Ther* 7: R217–R229

70   Katschke KJ Jr, Rottman JB, Ruth JH, Qin S, Wu L, LaRosa G, Ponath P, Park CC, Pope RM, Koch AE (2001) Differential expression of chemokine receptors on peripheral blood, synovial fluid and synovial tissue monocytes/macrophages in rheumatoid arthritis. *Arthritis Rheum* 44: 1022–1032

71   Borzi RM, Mazzetti I, Cattini L, Uguccioni M, Baggiolini M, Facchini A (2000) Human chondrocytes express functional chemokine receptors and release matrix-degrading enzymes in response to C-X-C and C-C chemokines. *Arthritis Rheum* 43: 1734–1741

72   Ruth JH, Rottman JB, Katschke KJ Jr, Qin S, Wu L, LaRosa G, Ponath P, Pope RM, Koch AE (2001) Selective lymphocyte chemokine receptor expression in the rheumatoid joint. *Arthritis Rheum* 44: 2750–2760

73   Prahalad S (2006) Negative association between the chemokine receptor CCR5-Δ32 polymorphism and rheumatoid arthritis: A metaanalysis. *Genes Immun* 7: 264–268

74   Galligan CL, Matsuyama W, Matsukawa A, Mizuta H, Hodge DR, Yoshimura T (2004) Up-regulated expression and activation of the orphan chemokine receptor, CCRL2, in rheumatoid arthritis. *Arthritis Rheum* 50: 1806–1814

75   Patterson AM, Siddall H, Chamberlain G, Gardner L, Middleton J (2002) Expression of the Duffy antigen/receptor for chemokines (DARC) by the inflamed synovial endothelium. *J Pathol* 197: 108–116

76   Mullazehi M, Mathsson L, Lampa J, Ronnelid J (2006) Surface-bound anti-type II collagen-containing immune complexes induce production of tumor necrosis factor alpha,

interleukin-1β and interleukin-8 from peripheral blood monocytes *via* Fcγ receptor IIA. *Arthritis Rheum* 54: 1759–1771

77   Rantapaa-Dahlqvist S, Boman K, Tarkowski A, Hallmans G (2007) Upregulation of monocyte chemoattractant protein-1 expression in anti-citrulline antibody and immuno-globulin M rheumatoid factor positive subjects precedes onset of inflammatory response and development of overt rheumatoid arthritis. *Ann Rheum Dis* 66: 121–123

78   Loetscher P, Dewald B, Baggiolini M, Seitz M (1994) Monocyte chemoattractant protein 1 and interleukin 8 production by rheumatoid synoviocytes: Effects of anti-rheumatic drugs. *Cytokine* 6: 162–170

79   Lopez-Armada MJ, Sanchez-Pernaute O, Largo R, Diez-Ortego I, Palacios I, Egido J, Herrero-Beaumont G (2002) Modulation of cell recruitment by anti-inflammatory agents in antigen-induced arthritis. *Ann Rheum Dis* 61: 1027–1030

80   Volin MV, Campbell PL, Connors MA, Woodruff DC, Koch AE (2002) The effect of sulfasalazine on rheumatoid arthritis synovial tissue chemokine production. *Exp Mol Pathol* 73: 84–92

81   Ho CY, Wong CK, Li EK, Tam LS, Lam CW (2003) Suppressive effect of combination treatment of leflunomide and methotrexate on chemokine expression in patients with rheumatoid arthritis. *Clin Exp Immunol* 133: 132–138

82   Ellingsen T, Hornung N, Moller BK, Poulsen JH, Stengaard-Pedersen K (2007) Differential effect of methotrexate on the increased CCR2 density on circulating CD4 T lymphocytes and monocytes in active chronic rheumatoid arthritis, with down regulation only on monocytes in responders. *Ann Rheum Dis* 66: 151–157

83   Taylor PC, Peters AM, Paleolog E, Chapman PT, Elliott MJ, McCloskey R, Feldmann M, Maini RN (2000) Reduction of chemokine levels and leukocyte traffic to joints by tumor necrosis factor alpha blockade in patients with rheumatoid arthritis. *Arthritis Rheum* 43: 38–47

84   Klimiuk PA, Sierakowski S, Domyslawska I, Chwiecko J (2006) Regulation of serum chemokines following infliximab therapy in patients with rheumatoid arthritis. *Clin Exp Rheumatol* 24: 529–533

85   Torikai E, Kageyama Y, Suzuki M, Ichikawa T, Nagano A (2006) The effect of infliximab in chemokines in patients with rheumatoid arthritis. *Clin Rheumatol* 26: 1088–1093

86   Kageyama Y, Torikai E, Nagano A (2006) Anti-tumor necrosis factor.alpha antibody treatment reduces serum CXCL16 levels in patients with rheumatoid arthritis. *Rheumatol Int* 27: 467–472

87   Nissinen R, Leirisalo-Repo M, Peltomaa R, Palosuo T, Vaarala A (2004) Cytokine and chemokine receptor profile of peripheral blood mononuclear cells during treatment with infliximab in patients with active rheumatoid arthritis. *Ann Rheum Dis* 63: 681–687

88   Aeberli D, Seitz M, Juni P, Villiger PM (2005) Increase of peripheral CXCR3 positive T lymphocytes upon treatment of RA patients with TNF-alpha inhibitors. *Rheumatology* (Oxford) 44: 172–175

89   van der Voort R, van Lieshout AW, Toonen LW Sloetjes AW, van den Berg WB, Figdor CG, Radstake TR, Adema GJ (2005) Elevated CXCL16 expression by synovial

macrophages recruits memory T cells into rheumatoid joints. *Arthritis Rheum* 52: 1381–1391

90 Newton SM, Mackie SL, Martineau AR, Wilkinson KA, Kampmann B, Fischer C, Dutta S, Levin M, Wilkinson RJ, Pasvol G (2008) Reduction of chemokine secretion in response to mycobacteria in infliximab-treated patients. *Clin Vaccine Immunol* 15: 506–512

91 Sato M, Miyazaki T, Nagaya T, Murata Y, Ida N, Maeda K, Seo H (1996) Antioxidants inhibit tumor necrosis factor-alpha mediated stimulation of interleukin-8, monocyte chemoattractant protein-1, and collagenase expression in cultured human synovial cells. *J Rheumatol* 23: 432–438

92 Yokota K, Miyazaki T, Hirano M, Akiyama Y, Mimura T (2006) Simvastatin inhibits production of interleukin-6 (IL-6) and IL-8 and cell proliferation induced by tumor necrosis factor-alpha in fibroblast-like synoviocytes from patients with rheumatoid arthritis. *J Rheumatol* 33: 463–471

93 Wang Y, Wei D, Lai Z, Le Y (2006) Triptolide inhibits CC chemokines expressed in ratd adjuvant-induced arthritis. *Int Immunopharmacol* 6: 1825–1832

94 Ahmed S, Pakozdi A, Koch AE (2006) Regulation of interleukin-1β-induced chemokine production and matrix metalloproteinase 2 activation by epigallocatechin-3-gallate in rheumatoid arthritis synovial fibroblasts. *Arthritis Rheum* 54: 2393–2401

95 Ospelt C, Kurowska/Stolarska M, Neidhart M, Michel BA, Gay RE, Laufer S, Gay S (2008) The dual inhibitor of lipoxygenase and cyclooxygenase ML3000 decreases the expression of CXCR3 ligands. *Ann Rheum Dis* 67: 524–529

96 Hounoki H, Sugiyama E, Mohamed SG, Shinoda K, Taki H, Abdel-Aziz HO, Maruyama M, Kobayashi M, Miyahara T (2008) Activation of peroxisome proliferator-activated receptor gamma inhibits TNF-alpha mediated osteoclast differentiation in human peripheral monocytes in part *via* suppression of monocyte chemoattractant protein-1 expression. *Bone* 42: 765–774

97 Akahoshi T, Endo H, Kondo H, Kashiwazaki S, Kasahara T, Mukaida N, Harada A, Matsushima K (1994) Essential involvement of interleukin-8 in neutrophil recruitment in rabbits with acute experimental arthritis induced by lipopolysaccharide and interleukin-1. *Lymphokine Cytokine Res* 13: 113–116

98 Halloran MM, Woods JM, Strieter RM, Szekanecz Z, Volin MV, Hosaka S, Haines GK III, Kunkel SL, Burdick MD, Walz A, Koch AE (1999) The role of an epithelial neutrophil-activating peptide-78-like protein in rat adjuvant-induced arthritis. *J Immunol* 162: 7492–7500

99 Kasama T, Strieter RM, Lukacs NW, Lincoln PL, Burdick MD, Kunkel SL (1995) Interleukin-10 expression and chemokine regulation during the evolution of murine type II collagen-induced arthritis. *J Clin Invest* 95: 2868–2876

100 Salomon I, Netzer N, Wildbaum G, Schif-Zuck S, Maor G, Karin N (2002) Targeting the function of IFNγ-inducible protein 10 suppresses ongoing adjuvant arthritis. *J Immunol* 169: 2865–2873

101 Skov L, Beurskens FJ, Zachariae CO, Reitamo S, Teeling J, Satijn D, Knudsen KM, Boot

EP, Hudson D, Baadsgaard O, Parren PW, van de Winkel JG (2008) IL-8 as antibody therapeutic target in inflammatory diseases: Reduction of clinical activity in palmoplantar pustulosis. *J Immunol* 181: 669–679

102 Chintalacharuvu SR, Wang JX, Giaconia JM, Venkataraman C (2005) An essential role for CCL3 in the development of collagen-induced arthritis. *Immunol Lett* 100: 202–204

103 Ogata H, Takeya M, Yoshimura T, Takagi K, Takahashi K (1997) The role of monocyte chemoattractant protein-1 (MCP-1) in the pathogenesis of collagen-induced arthritis in rats. *J Pathol* 182: 106–114

104 Schrier DJ, Schimmer RC, Flory CM, Tung DK, Ward PA (1998) Role of chemokines and cytokines in a reactivation model of arthritis in rats induced by injection with streptococcal cell walls. *Leukoc Biol* 63: 359–363

105 Gong JH, Ratkay LG, Waterfield JD, Clark-Lewis I (1997) An antagonist of monocyte chemoattractant protein 1 (MCP-1) inhibits arthritis in the MRL-lpr mouse model. *J Exp Med* 186: 131–137

106 Barnes DA, Tse J, Kaufhold M, Owen M, Hesselgesser J, Strieter R, Horuk R, Perez HD (1998) Polyclonal antibody directed against human RANTES ameliorates disease in the Lewis rat adjuvant-induced arthritis model. *J Clin Invest* 101: 2910–2919

107 Inoue T, Yamashita M, Higaki M (2001) The new antirheumatic drug KE-298 suppresses MCP-1 and RANTES production in rats with adjuvant-induced arthritis and in IL-1–stimulated synoviocytes of patients with rheumatoid arthritis. *Rheumatol Int* 20: 149–153

108 Haringman JJ, Gerlag DM, Smeets TJ, Baeten D, van den Bosch F, Bresnihan B, Breedveld FC, Dinant HJ, Legay F, Gram H et al (2006) A randomized, controlled trial with an anti-CCL2 (anti-monocyte chemotactic protein 1) monoclonal antibody in patients with rheumatoid arthritis. *Arthritis Rheum* 54: 2387–2392

109 Nanki T, Urasaki Y, Imai T, Nishimura M, Muramoto K, Kubota T, Miyasaka N (2004) Inhibition of fractalkine ameliorates murine collagen-induced arthritis. *J Immunol* 173: 7010–7016

110 Bursill CA, Cai S, Channon KM, Greaves DR (2003) Adenoviral-mediated delivery of a viral chemokine binding protein blocks CC-chemokine activity *in vitro* and *in vivo*. *Immunobiology* 207: 187–196

111 Gong JH, Yan R, Waterfield JD, Clark-Lewis I (2004) Post-onset inhibition of murine arthritis using combined chemokine antagonist therapy. *Rheumatology* 43: 39–42

112 Youssef S, Maor G, Wildbaum G, Grabie N, Gour-Lavie A, Karin N (2000) C-C chemokine-encoding DNA vaccines enhance breakdown of tolerance to their gene products and treat ongoing adjuvant arthritis. *J Clin Invest* 106: 361–371

113 Matsukawa A, Yoshimura T, Fujiwara K, Maeda T, Ohkawara S, Yoshinaga M (1999) Involvement of growth-related protein in LPS-induced rabbit arthritis. *Lab Invest* 79: 591–600

114 Podolin PL, Bolognese BJ, Foley JJ, Schmidt DB, Buckley PT, Widdowson KL, Jin Q, White JR, Lee JM, Goodman RB et al (2002) A potent and selective nonpeptide antago-

nist of CXCR2 inhibits acute and chronic models of arthritis in the rabbit. *J Immunol* 169: 6435–6444

115 Cunha TM, Barsante MM, Guerrero AT, Verri WA jr, Ferreira SH, Coelho FM, Bertini R, Di Giacinto C, Allegretti M, Cunha FQ, Teixeira MM (2008) Treatment with DF 2162, a non-competitive allosteric inhibitor of CXCR1/2, diminishes neutrophil influx and inflammatory hypernociception in mice. *Br J Pharmacol* 154: 460–470

116 Brown CR, Blaho VA, Loiacono CM (2003) Susceptibility to experimental Lyme arthritis correlates with KC and monocyte chemoattractant protein-1 production in joints and requires neutrophil recruitment *via* CXCR2. *J Immunol* 171: 893–901

117 Gao P, Zhou XY, Yashiro-Ohtani Y, Yang YF, Sugimoto N, Ono S, Nakanishi T, Obika S, Imanishi T, Egawa T et al (2003) The unique target specificity of a nonpeptide chemokine receptor antagonist: Selective blockade of two Th1 chemokine receptors CCR5 and CXCR3. *J Leukoc Biol* 73: 273–280

118 Hatse S, Princen K, Bridger G, DeClerq E, Schols D (2002) Chemokine receptor inhibition by AMD3100 is strictly confined to CXCR4. *FEBS Lett* 527: 255–262

119 Tamamura H, Fujii N (2005) The therapeutic potential of CXCR4 antagonists in the treatment of HIV infection, cancer metastasis and rheumatoid arthritis. *Expert Opin Ther Targets* 9: 1267–1282

120 Pease JR, Horuk R (2005) CCR1 antagonists in clinical development. *Expert Opin Investig Drugs* 14: 785–796

121 Gladue RP, Tylaska LA, Brissette WH, Lira PD, Kath JC, Poss CS, Brown MF, Paradis TJ, Conklyn MJ, Ogborne KT (2003) CP-481,715, a potent and selective CCR1 antagonist with potential therapeutic implications for inflammatory diseases. *J Biol Chem* 278: 40473–40480

122 Haringman JJ, Kraan MC, Smeets TJM, Zwinderman KH, Tak PP (2003) Chemokine blockade and chronic inflammatory disease: Proof of concept in patients with rheumatoid arthritis. *Ann Rheum Dis* 62: 715–721

123 Quinones MP, Estrada CA, Kalkonde Y, Ahuja SK, Kuziel WA, Mack M, Ahuja SS (2005) The complex role of the chemokine receptor CCR2 in collagen-induced arthritis: Implications for therapeutic targeting of CCR2 in rheumatoid arthritis. *J Mol Med* 83: 672–681

124 Quinones MP, Jimenez F, Martinez H, Estrada CA, Willmon O, Dudley M, Kuziel WA, Melby PC, Reddick RL, Ahuja SK et al (2006) CC chemokine receptor (CCR)-2 prevents arthritis development following infection by *Mycobacterium avium*. *J Mol Med* 84: 503–512

125 Amat M, Benjamim CF, Williams LM, Prats N, Terricabras E, Beleta J, Kunkel SL, Godessart N (2006) Pharmacological blockade of CCR1 ameliorates murine arthritis and alters cytokine networks *in vivo*. *Br J Pharmacol* 149: 666–675

126 Plater-Zyberk C, Hoogewerf AJ, Proudfoot AE, Power CA, Wells TN (1997) Effect of a CC chemokine receptor antagonist on collagen induced arthritis in DBA/1 mice. *Immunol Lett* 57: 117–120

127 Shahrara S, Proudfoot AE, Woods JM Ruth JH, Amin MA, Park CC, Haas CS, Pope

RM, Haines GK, Zha YY et al (2005)Amelioration of rat adjuvant-induced arthritis by Met-RANTES. *Arthritis Rheum* 52: 1907–1919

128 Yang YF, Mukai T, Gao P Yamaguchi N, Ono S, Iwaki H, Obika S, Imanishi T, Tsujimura T, Hamaoka T et al (2002) A non-peptide CCR5 antagonist inhibits collagen-induced arthritis by modulating T cell migration without affecting any collagen T cell responses. *Eur J Immunol* 32: 2124–2132

129 Brühl H, Cihak J, Plachy J, Kunz-Schughart L, Niedermeier M, Denzel A, Rodriguez-Gomez M, Talke Y, Luckow B, Stangassinger M, Mack M (2007) Targeting of Gr-1[+], CCR2[+] monocytes in collagen-induced arthritis. *Arthritis Rheum* 56: 2975–2985

130 Quinones MP, Ahuja SK, Jimenez F Schaefer J, Garavito E, Rao A, Chenaux G, Reddick RL, Kuziel WA, Ahuja SS (2004) Experimental arthritis in CC chemokine receptor 2-null mice closely mimics severe human rheumatoid arthritis. *J Clin Invest* 113: 856–866

131 Bao L, Zhu Y, Zhu J, Lindgren JU (2005) Decreased IgG production but increased MIP-1β expression in collagen-induced arthritis in C-C chemokine receptor 5-deficient mice. *Cytokine* 31: 64–71

132 Clucas AT, Shah A, Zhang YD, Chow VF, Gladue RP (2007) Phase I evaluation of the safety, pharmacokinetics and pharmacodynamics of CP-481,715. *Clin Pharmacokinet* 46: 757–766

133 Vergunst CE, Gerlag DM, Lopatinskaya L, Klareskog L, Smith MD, van den Bosch F, Dinant HJ, Lee Y, Wyant T, Jacobson EW et al (2008) Modulation of CCR2 in rheumatoid arthritis: A double-blind, randomized, placebo-controlled clinical trial. *Arthritis Rheum* 58: 1931–1939

134 Meanwell NA, Kadow JF (2007) Maraviroc, a chemokine CCR5 receptor antagonist for the treatment of HIV infection and AIDS. *Curr Opin Investig Drugs* 8: 669–681

Signaling pathways in rheumatoid arthritis

Jean-Marc Waldburger[1,2] and Gary S. Firestein[1]

[1]University of California, San Diego, Department of Medicine, La Jolla, CA, 92093-0656, USA
[2]University of Geneva, Centre médical universitaire, 1 rue Michel Servet, 1211 Geneva, Switzerland

Abstract

Signaling pathways orchestrate the inflammatory response by regulating various cellular functions such as programmed cell death, cell differentiation and proliferation or secretion of signaling molecules. They are classically activated by ligand engagement of surface receptors but increasing evidence suggests that intracellular proteins can also detect danger signals. Protection from pathogens, chemical or physical injury, or neoplasia relies on a tightly regulated activation of these mechanisms. The same signaling cascades sometimes escape from normal controls and increase the production of cytokines, proteases, growth factors and chemokines up to harmful levels, leading to an autodestructive process as seen in rheumatoid arthritis. Mapping the hierarchy of these pathways identifies which specific targets can be inhibited to safely reduce the levels of inflammatory molecules and reset the homeostasis of the organism.

Introduction

Intracellular signaling pathways provide cells with the ability to respond to extracellular signals in their environment. Depending on the nature of the stimulus and the particular cell type involved, the signaling pathways regulate various cellular functions such as programmed cell death, cell growth, differentiation and proliferation. Protection from pathogens, chemical or physical injury, or neoplasia also involves intracellular pathways that regulate the expression of cytokines, chemokines and other secreted or membrane-bound molecules that contribute to host defense. When signaling cascades escape from normal controls, they can lead to a pathological or destructive process as seen in diseases like systemic lupus erythematosus (SLE) or rheumatoid arthritis (RA).

Many extra- and intracellular mediators are released into the environment during inflammation. These molecules arise from endogenous sources (hormones, cytokines, free radicals, products of cell metabolism or apoptosis) or from exogenous microorganisms. The most common methods for activating intracellular signaling cascades include ligand engagement of surface receptors, although increasing evidence suggests that intracellular receptors play a key role in detecting danger signals. Other signal transduction pathways are activated by physical or chemical stressors, such as heat, ultraviolet radiation, or osmotic changes.

The complex signaling mechanisms involved in rheumatic diseases like RA converge on key pathways including the mitogen-activated protein kinase (MAPK) and nuclear factor-κB (NF-κB) pathways. In each case, transcription factors that control the production of cytokines, proteases, growth factors and chemokines are activated. Post-transcriptional and post-translational events further regulate mediator production. The careful study of these signal transduction steps can potentially identify novel therapeutic targets to interrupt uncontrolled host responses that damage the host.

Upstream components of intracellular pathways in RA

The key cytokines interleukin-1 (IL-1) and tumor necrosis factor (TNF) are rapidly released by cells of the innate immune system as a first line of defense against exogenous pathogens. During chronic autoimmune diseases such as RA, cytokines orchestrate tissue injury and are proposed to maintain an autoinflammatory loop. The clinical success of biological agents in RA underlines the clinical importance of TNF and, to a lesser extent IL-1. However, the available anti-cytokine therapies induce a true remission in only a minority of patients, which is probably due to the redundancy of the signaling network in RA and the heterogeneous nature of the disease.

IL-1 signaling

The IL-1 family of cytokines comprises 11 different members as defined by amino acid sequence homologies. Several of these proteins are thought to play important roles in the pathology of RA, including IL-1β, IL-1α and IL-18 [1]. IL-1β needs cleavage of its pro-form by the cysteine protease caspase-1 to gain biological activity. In contrast, both pro- and mature IL-1α can induce cellular responses. In RA, monocytes/macrophages are the principal IL-1-producing cells. IL-1 stimulates the expression of multiple cytokines and inflammatory factors that drive extracellular matrix degradation.

Several animal models confirm the dominant role of IL-1 during joint inflammation. IL-1-deficient mice are resistant to experimental arthritis [2]. Conversely, the absence of the natural IL-1 receptor antagonist (IL-1RA) exposes the animals to a spontaneous autoimmune syndrome that mainly targets the joints [3]. One agent that specifically blocks IL-1 signaling is a recombinant version of IL-1RA (anakinra) [4]. While randomized trials clearly prove modest benefit of this agent in RA, the overall clinical experience suggests that anakinra is less effective than TNF inhibitors. This contrasts with the remarkable effectiveness of anakinra in

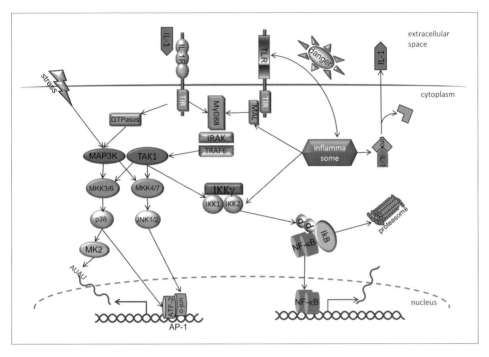

Figure 1
Signaling pathways for IL-1 and Toll-like receptors (TLRs). Activation of IL-1 receptors or TLRs, such as TLR4 by lipopolysaccharide, converge on MyD88 to activate NF-κB and the mitogen-activated protein kinases (MAPKs) like p38 and JNK. These lead to transcription of genes involved in the inflammatory response. The inflammasome is also activated by environmental danger signals, thereby activating IL-1 cleavage and NF-κB.

the management of other systemic inflammatory disorders including systemic onset juvenile idiopathic arthritis or adult onset Still's disease. Because Toll-like receptors (TLR) and IL-1 share a signaling pathway through the myeloid differentiation primary response gene (MyD88) (see Fig. 1), the protective effect of blocking IL-1 in animal models of RA can be overcome by small doses of TLR ligands [5]. Thus, the presence of alternative pathways to IL-1 signaling might explain the relative low potency of anakinra in RA, which has also been linked to pharmacological limits of the drug including a short half-life.

IL-1 is a potent inflammatory mediator. It is therefore not surprising that its activity and production are tightly regulated at many different steps including transcriptional and translational control, processing of pro-IL-1 by the inflammasome (see below), a receptor antagonist (IL-1RA), a decoy receptor (IL-1RII), the IL-1 receptor accessory protein (IL-1RAcP) and finally modulation of intracellular signaling pathways. The IL-1R complex comprises IL-1RI and IL-1RAcP. The sole known function of IL-1RII is to quench the activity of IL-1 [6]. Both the membrane-bound and the soluble form of IL-1RII are able to capture IL-1. The cytoplasmic tail of IL-1 RII lacks a TIR domain and cannot transduce intracellular signaling [7]. Moreover, IL-1RII can form a dominant negative complex with IL-1RAcP. IL-1RA acts as a classical receptor antagonist because of its higher affinity for IL-1RI compared with IL-1 itself. Because IL-1RII binds IL-1RA much less efficiently than IL-1RI, the two anti-IL-1 molecules synergize.

The IL-1R–IL-1 complex is recognized by IL-1RAcP. The result of this association is the formation of a heterodimeric transmembrane receptor complex where both IL-1R and IL-1RAcP are needed to initiate signal transduction. The close spatial association of IL-1R and IL-1RAcP in the complex presumably allows homotypic protein-protein interactions of their respective TIR domains. Conformational changes result that enable recruitment of the cytosolic TIR domain-containing adaptor protein MyD88. MyD88 then interacts with IL-1R-associated kinases (IRAKs), leading to the activation and ubiquitinylation of TNF receptor-associated factor 6 (TRAF6). Downstream of TRAF6, TAK1, a mitogen-activated protein (MAP) 3 kinase, activates the IκB kinase (IKK) complex and thus NF-κB, leading to the expression of inflammatory cytokine genes. In addition, TAK1 with other MAP3Ks launches MAPK activation, which further increase cytokine levels.

Endogenous regulatory molecules have been described along the IL-1R intracellular pathway. MyD88s, a splice variant of MyD88, lacks the interaction domain involved in IRAK4 recruitment and acts as a natural dominant negative molecule that inhibits NF-κB activation [8]. IRAK-M prevents the dissociation of IRAK1 and IRAK4 from MyD88 [9].

To minimize potential infectious side effects, small molecules that specifically target IL-1R signaling could be designed to leave other TLR responses needed for host defense intact. In RA, however, it might be desirable to block some of the other TLR pathways as well, since many might be involved in the pathogenesis of synovial inflammation (see below). Direct targeting of most signaling components downstream of IL-1R will affect other TLRs because they share common kinases and adaptors (except TLR3, which does not use MyD88). Another strategy relies on disrupting TIR-TIR domain homotypic contacts. The interaction between MyD88 and IL-1RI can be disrupted by a small molecule [10]. This compound showed no inhibitory effect on the association of TLR4 with Myd88 and was thus able to interfere with IL-1 signaling but left LPS responses intact.

TNF

In RA, the efficacy of TNF inhibitors currently approved for clinical use correlates with animal models proving TNF as a major regulator of joint inflammation and destruction [11]. TNF is the prototype of a superfamily of ligands with a common trimeric structure, which is their biologically active form, either as a membrane-anchored protein or in their soluble form [12]. TNF cytokine family members interact with a group of cognate receptors (TNFR) and most ligands interact with more than one receptor. TNFR1 and TNFR2 both interact with the homotrimers of lymphotoxin-α (LT-α) and TNF. Signal transmission by TNF is initiated by ligand-induced clustering of receptors. TNF has various and opposing functions depending on the cellular context, such as apoptosis, differentiation, proliferation or expression of inflammatory genes through the activation of pathways involving NF-κB and MAPK.

None of the receptors of the mammalian TNF superfamily has intrinsic enzymatic activity. A basic framework has been assembled to account for the seeming ambivalence of activation by which TNFR can assemble signaling complexes for both the caspase-8 apoptotic and the NF-κB anti-apoptotic pathways (see Fig. 2). TNFR1

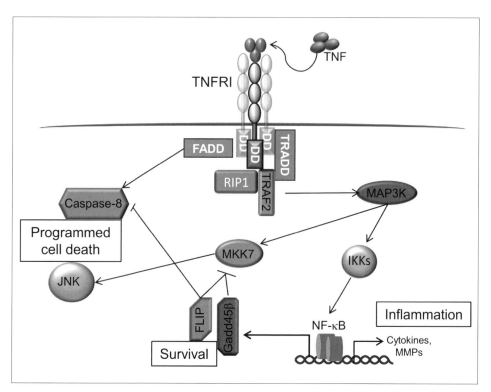

Figure 2
Signaling pathways for TNF. The TNF receptor assembles a signaling complex that can activate the MAPKs, NF-κB, and caspases. This process can induce transcription of pro-inflammatory genes and influence cell survival.

couples to the apoptotic cascade by recruiting the adaptors TRADD (TNFR-associated death domain) and FADD (Fas-associated death domain). TNFR1 and TNFR2 can also engages TNFR-associated factors (TRAFs) to couple to NF-κB activation, both a regulator of apoptosis and a major inducer of inflammatory genes. Resistance to apoptosis after TNF signaling requires *de novo* protein synthesis of the FLICE inhibitory protein (FLIP) and Gadd45β, whose expressions are dependent on NF-κB [13]. Prolonged activation of c-Jun-N-terminal kinase (JNK) by TNF causes proteolysis of FLIP. Gadd45β binds to and inactivates MKK7, an upstream activating kinase of JNK, and protects from TNF-induced apoptosis [14]. These divergent actions explain why the apoptotic pathway dominates in cells that are biosynthetically compromised. In RA, TNF activates mostly inflammatory genes.

Pattern-recognition receptor signaling

Pattern recognition receptors (PRRs) alert the host by detecting pathogen-associated molecular patterns (PAMPs) during infection. Several PRRs also recognize self molecules especially those released from dying or damaged cells. By itself, activation of these receptors triggers rapid inflammatory reactions during innate immune responses. Moreover, they prime antigen-presenting cells (APCs) to engage the adaptive immune system. These two properties are central to their role in host defense and homeostasis but can also lead to pathological outcomes. Inappropriate activation of these receptors by self molecules, commensal bacteria, or pathogens during acute or chronic infection has been associated with autoimmune diseases such as RA. PRRs are classified into membrane-bound and cytoplasmic receptors. TLRs are the best-characterized cell surface membrane-associated PRRs. Cytoplasmic PRRs comprise CARD helicases and NOD-like receptors. Extracellular molecules such as the complement system or pentraxins (SAP, CRP) can also be classified as PRRs.

Cell surface membrane-associated PRRs: TLRs

TLRs are expressed on synovial fibroblast and infiltrating myeloid cells in human RA joints. Classical animal models such as adjuvant arthritis or streptococcal cell wall arthritis are dependent on the activation of the innate immune system by TLR ligands. The extent to which TLRs engage in chronic inflammation during RA is still unclear. TLR ligands of microbial and endogenous origin have been detected in joints of RA patients. *In vitro*, they are able to stimulate inflammatory cytokine production by synoviocytes. TLR-stimulated APCs can prime T cell responses towards aggressive phenotypes such as Th1 or Th17 depending on the ligand and APC subtype, thus providing a link between the innate and the adaptive immune system. In B cells, TLRs influence antibody class switching and rheumatoid factor (RF) production.

The different TLRs initiate both shared and distinct signaling pathways by recruiting four different TIR domain-containing adaptor molecules (see Fig. 1): MyD88, TIRAP (MAL), TRIF (TICAM1), and TRAM [15]. MyD88 is used by all TLRs (and IL-1R, see above) except TLR3; TIRAP is downstream of TLR2 and TLR4; TRIF is recruited by TLR3 and TLR4; and TRAM is used only by TLR4. These signaling pathways raise inflammatory cytokine production *via* the transcription factors NF-κB, activator protein-1 (AP-1) and MAPKs such as JNKs and p38. TLRs 3, 4, 7, 8, and 9 also activate interferon (IFN) regulatory factor 3 (IRF3) and/ or IRF7, leading to the production of type I IFNs such as IFN-β and IFN-α.

TLR4 is expressed in RA joints and its ligand LPS can trigger arthritis in several rodent models [16]. The serum and synovial fluid of RA patients contain an unknown TLR4 ligand [17]. TLR4 sets up both MyD88-dependent and TRIF-dependent pathways. The MyD88-dependent pathway is shared with IL-1 and induces the expression of inflammatory cytokines such as IL-6 in fibroblasts and IL-12 and TNF-α in myeloid cells. The MyD88-independent pathway involves the consecutive recruitment of TRAM, TRIF then TRAF6, which links to TAK1 and other MAP3Ks activation. Downstream of TRIF, RIP1 also connects to NF-κB and to TRAF3, which evokes IRF3-dependent antiviral responses. The TRIF and MyD88 pathways synergize to maximize the expression of inflammatory cytokines by providing optimal MAPK and NF-κB activation.

RNA released from necrotic synovial fluid cells can activate rheumatoid fibroblast to produce high levels of IFN-β, CXCL10, CCL5, and IL-6 protein by binding to TLR3 [18]. TLR3 shares the TRIF-TRAF3 pathway of TLR4, which eventually triggers the IRF3-dependent antiviral program *via* TBK1 and IKK. TLR3 also signals to NF-κB through TRAF6 and TAK1, which might account for induction of IL-6 in synovial fibroblasts exposed to RNA or poly(I:C). TLR3 ligands increase IRF-3 phosphorylation and production of IFN and RANTES in cultured synoviocytes, *via* IKKε [19]. Activation of both IRF3 and c-Jun by IKKε also increases matrix metalloproteases (MMP) expression in synovial tissue [20]. The dual role of IKKε in the synovial inflammatory response suggests that this pathway is a potential therapeutic target in arthritis.

In RA, TLR2 is chiefly expressed in fibroblast-like synoviocytes (FLS) and infiltrating lymphocytes [16]. Stimulation of FLS with the TLR2 ligand peptidoglycan but not the TLR9 ligand CpG up-regulates the expression of IL-6, IL-8, chemokines and MMPs 1, 3, and 13 [21]. TLR2 can broaden the repertoire of its ligands and the outcome of its signaling by collaborating with other TLRs or non-TLR PRRs. This is relevant to the clinical setting where the response to PRR is caused by complex ligands as opposed to agonists chemically purified *in vitro*.

TLR9 and TLR7 are thought to link innate recognition of endogenous DNA and RNA, respectively, to pathological B and dendritic cell activation in autoimmune diseases. Immune complexes of IgG and self DNA induce RF production by activating autoreactive B cells [22]. The signal triggered by TLR9 synergizes with the RF B cell receptor, which is activated by the Fc portion of IgGs in the immune complex. TLR9/7 stimulation also leads to type I IFN production. Antiviral gene products have been shown to be involved in the pathology of SLE but might be protective in arthritis models. TLR9/7 stimulates the synthesis of inflammatory cytokines and intra-articular injection of bacterial DNA containing unmethylated CpG motifs causes arthritis [23]. The inflammatory pathway downstream of TLR9/7 is similar to TLR4 (MyD88-IRAK4-TRAF6-TAK1-MAPK/IKK-AP-1/NF-κB). On the other hand, type I IFN production differs from TLR3 and TLR4 signaling since TRIF and IRF3 are dispensable [24]. Instead, a multiprotein complex comprising at least MyD88, IRAK4, IRAK1, TRAF6, and TRAF3 activates the antiviral transcription factor IRF7.

Cytoplasmic PRRs: NLRs

Soluble cytoplasmic PRRs include nucleotide-binding oligomerization domain (NOD)-like receptors (NLRs), double-stranded RNA-activated protein kinase (PKR) and caspase activation and recruitment domain (CARD) helicases. Mammalian NLR family members share a modular organization of their protein structure that consists of a C-terminal leucine-rich repeat (LRR) domain, a central nucleotide-binding NACHT domain and an N-terminal protein-protein–interaction domain composed of a CARD domain, pyrin domain or Bir (baculovirus 'inhibitor of apoptosis' repeat) domain [25]. Only 3 human NOD-LRR proteins have N-terminal CARDs, whereas NOD-LRR members with a PYD include 14 proteins named NALPs. Some NLRs mutations predispose to inflammatory diseases. A prominent example is the genetic association of CIAS-1 (the gene that codes for cryopyrin) with three autoinflammatory diseases: familial cold autoinflammatory syndrome, Muckle-Wells syndrome, and neonatal onset multisystem inflammatory disease [26].

The PRR role of NALP1, NALP3 (NACHT-, LRR- and pyrin-domain-containing proteins) and IPAF (ICE-protease-activating factor) is controlled by a cytosolic multiprotein complex called the inflammasome (see Fig. 1). This platform couples the sensing of cytoplasmic PAMP to the proteolysis of pro-IL-1β into its bioactive 17-kDa fragment. Each NALP senses distinct PAMPs and danger signals. On ligand recognition, the NALP cleaves pro-caspase-1 into active caspase-1 or ICE (IL-1β converting enzyme). This usually needs the adaptor ASC (apoptosis-associated speck-like protein containing a CARD). IPAF has been shown to activate caspase-1 without ASC.

PAMPs need to reach the cytosol to activate NLRs. The secretion machinery of some bacteria can deliver their PAMPs straight in the cytosol. Ionophores such as nigericin or extracellular ATP cause potassium efflux, which activates the inflammasome by a pannexin-1-dependent mechanism. This hemichannel molecule forms a large pore after intracellular potassium depletion through which extracellular PAMPs gain access to the cytosol [27]. The purinergic receptor P2X7 mediates ATP-dependent potassium loss. How NLR family members sense their agonists in the cytosol remains poorly understood. The demonstration of a direct physical interaction between NLRs and microbial components or endogenous danger molecules is still lacking and might involve another set of unknown adaptors.

NLR PRRs can also signal to NF-κB. NOD1 and NOD2 ligands are the peptidoglycan (PGN)-derived peptides γ-D-glutamyl-meso-diaminopimelic acid (iE-DAP) and MDP, respectively. NOD1 and NOD2 can form a complex with RIP2 (receptor-interacting protein 2), which leads to activation of NF-κB. Caspase-1 can activate Mal (TIRAP) to link NF-κB and p38 MAP kinase pathways [28]. Mutations in both Card4 (NOD1) and Card15 (NOD2) can cause inflammatory diseases. Variants linked to Crohn's disease (CD) are mapped to the LRR region of NOD2, whereas those encoded in the NACHT domain are found in Blau syndrome and early onset sarcoidosis patients. Many of these polymorphisms confer a gain of function to NOD2 and increase NF-κB activation [29].

Beside the many regulatory mechanisms of IL-1 signaling, endogenous inhibitors of caspase-1 and therefore IL-1 release have been described. Caspase-12 has a dominant negative effect on caspase-1 [30]. CARD-containing molecules have been shown to bind to the CARD domain of pro-casapase-1 and to inhibit caspase-1 [31]. Pyrin, the protein mutated in familial Mediterranean fever patients (FMF), regulates caspase-1 activation by interacting with the adaptor ASC to inhibit caspase-1 [32].

Inhibition of the inflammasome is a new therapeutic alternative in IL-1-mediated diseases, which now also comprises crystal-induced arthropathies. P2X7 antagonists, which are under development for the treatment of multiple sclerosis, can inhibit the ATP pathway. Blocking Pannexin-1 could have a broader anti-IL-1 activity assuming a general role of this molecule in mediating cytosolic entry of extracellular NALP ligands. Pannexin-1 inhibitory peptides show efficacy *in vitro* [27]. The mechanism by which IL-1 induces itself is thought to be a major pathogenic loop in several inflammatory diseases and represents an unresolved issue [33]. Recombinant IL-1 antagonist is an effective therapy in cryopyrin-associated periodic syndromes (CAPS) [34]. This contrasts with the preventing action of colchicine in most but not all FMF variants [35]. Caspase-1 inhibitors, which are in clinical development, have the ability to block IL-1 and IL-18 processing regardless of the specificity of the NLR involved and can also block NF-κB activation by caspase-1 [36].

T cell receptor signaling in CD4+ T cells

The role of CD4+ T cells in the pathogenesis of RA is supported by recent data showing the efficacy of abatacept, which blocks T cell costimulation [37]. Many rodent models of RA are also T cell dependent at least in their initiation phase. Beside traditional models like collagen-induced arthritis, mice homozygous for a mutation in the gene encoding ZAP-70, a key signal transduction molecule in T cells, develop a T cell-mediated autoimmune arthritis [38].

T cell receptor (TCR)-mediated signaling pathways that might take part in rheumatic diseases are complex (see Fig. 3). Quantitative and qualitative modulation of

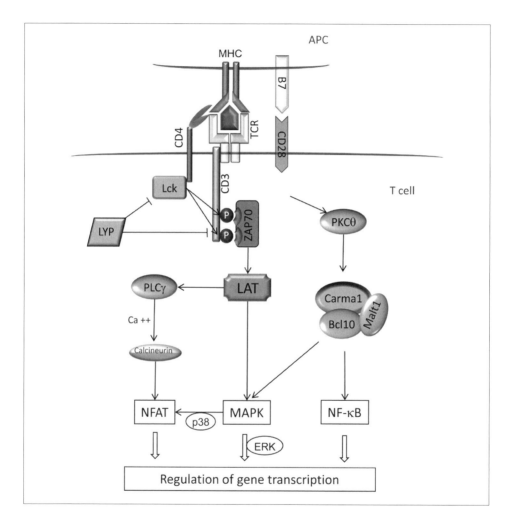

Figure 3
Signaling pathways for the T cell receptor (TCR). The TCR is activated by antigen in the context of MHC and engagement of costimulatory signals. ZAP70 leads to a cascade of signaling events, including NF-κB translocation and stimulation of the MAPK pathway. Depending on the intensity of the signal and the cytokine milieu, T cells can be directed along specific differentiation and activation pathways.

these pathways during antigenic stimulation can produce two main opposing outcomes: differentiation to effector T cells or anergy. Full activation of T cells depends on the engagement of the TCR and CD3 by the peptide-loaded MHC on the APC (signal I) and the costimulatory receptors CD28 and B7 (signal II). Cytokine production, survival and proliferation of the T cell need both signals, whereas engagement of the TCR without signal II induces anergy resulting in antigen tolerance even after rechallenge with proper costimulation.

The α and β chains of the TCR that recognize the antigenic peptide lack enzymatic activity and have a short cytoplasmic tail [39]. They are associated with the CD3 complex, the most upstream substrate of intracellular signaling kinases. On TCR triggering the src-family kinases Lck and Fyn phosphorylate two tyrosine residues within the immunoreceptor tyrosine-based activation motifs (ITAMs) of the CD3-ζ, -γ, -δ and -ε subunits. Phosphorylated ITAMs recruits ZAP-70 (ζ-chain-associated protein of 70 kDa). After it is attached to the ITAM motifs, Lck activates ZAP-70, the function of which is further increased by dephosphorylation of inhibitory sites. Higher affinity TCR-MHC contacts correlate with greater number of phosphorylated ITAM motifs up to the maximum of ten ITAMs present in each TCR complex. Downstream signaling events include the activation of Ras and Rho-family GTPases, MAPK cascades, phosphatidylinositol 3 kinase (PI3K), PKCθ, and the NF-κB pathway. The transcription of biologically important genes is also modulated by intracellular calcium flux triggered by phospholipase C (PLC)-γ.

Phosphorylation of the transmembrane adaptor molecule LAT and the cytosolic adaptor protein SLP-76 by ZAP-70 orchestrate these downstream events. Phosphorylated LAT and SLP-76 then serve as docking surfaces for other adaptors such as growth factor receptor-bound protein 2 (Grb2), GADS, the p85 regulatory subunit of PI3K and PLC-γ. Through these molecules, LAT effectively controls calcium flux, Ras/extracellular signal-regulated kinase (ERK), NFAT/AP-1, and PI3K activation. The activity of nuclear factor of activated T cells (NFAT), which promotes IL-2 transcription, is controlled by the sustained intracellular calcium flux initiated by PLC-γ. Ca^{2+} activates the serine phosphatase calcineurin, promoting the dephosphorylation and nuclear translocation of NFAT. Calcineurin is targeted by cyclosporine A and FK506 [39]. Binding of the adaptor Grb2 on LAT further recruits the GDP/GTP exchange factor Sos that activates the Ras/Raf/ERK pathway.

Once activated, the ERKs play an essential role in the expression of the AP-1 transcription factor c-Fos, as well as c-myc. c-Fos contributes to the transcriptional regulation of AP-1 response elements in the IL-2 promoter. Vav is recruited both by SLP-76 and Grb2 and is involved in NF-κB, NFAT, AP-1, and JNK activation and for sustained Ca^{2+} signaling. Vav and Lck are required for the membrane relocalization of PKCθ into lipid rafts where various other molecules, such as Lck, ZAP-70, LAT have been shown to accumulate [40]. The presence of PKCθ in these structures is necessary for NF-κB activation.

The T cell-dependent arthritis in the ZAP-70 mouse is mapped to a spontaneous homozygous missense mutation in the ZAP-70 gene [38]. The tyrosine-phosphorylation status of major signal transduction molecules, such as ZAP-70, TCR-ζ, LAT and PLC-γ is extremely low in T cells from SKG mice. This impairs most major signaling pathways including calcium flux, ERK, p38 and JNK, which raises the avidity threshold of the TCR. This causes decreased negative selection in the thymus because this process relies on the stronger TCR signals triggered by high avidity self-antigens-MHC complexes. Pathogenic self-reactive clones then accumulate in the periphery.

In RA, the most significant genetic association other than the HLA-DR is a polymorphism in the PTPN22 gene [41]. This variant confers a gain-of-function phenotype to its protein product, the LYP phosphatase, leading to decrease in tyrosine phosphorylation of proximal signaling molecules similar to the SKG mice [42]. By analogy, this PTPN22 SNP could thus lead to increased thymic output of autoreactive T cells in the periphery. However, LYP is also expressed in B cells, natural killer cells, macrophages, monocytes and dendritic cells suggesting that alterations in other cell types could participate in PTPN22-linked diseases.

Drugs designed to target proximal TCR events could replicate the autoreactive risk conveyed by the ZAP-70 mutation in mice or the PTPN22 SNPs in humans. In addition to skewing thymic selection, weakening proximal TCR signaling can interfere with the immunosuppressive regulatory T cells (Tregs). Tregs are implicated in animal models of arthritis [43]. They typically engage in a high-affinity interaction with self-peptide MHC complexes, which generate strong TCR signals. Superagonistic monoclonal antibodies with specificity for CD28 expand Tregs *in vitro* and *in vivo* and prove to be therapeutic in rat adjuvant arthritis [44]. However the role of this T cell subset in RA is not well defined and extrapolating their function from rodent models should be done with caution. In a first-in-man clinical trial, the CD28 superagonist antibody TGN1412 produced a severe inflammatory reaction resulting in multi-organ failure [45]. One potential explanation is that in humans TGN1412 stimulated the more abundant memory T cell subset, which overwhelmed any beneficial effects on Tregs. Another molecule intimately involved with TCR signaling, Janus kinase-3 (JAK3), has recently been evaluated using a small molecule inhibitor CP-690550. This compound demonstrated impressive clinical efficacy in Phase II clinical trials in RA. These proof of concept studies support the hypothesis that T cells participate in chronic rheumatoid synovitis, although effects on host defense are possible.

The therapeutic potential of the B7–CD28 costimulatory pathway in RA is demonstrated by the clinical success of abatacept, a soluble fusion protein containing the extracellular domain of CTLA-4 linked to an IgG Fc region [37]. It binds to B7 on the APC with greater affinity than CD28 on the T cell, blocking the so-called signal II that is required for the expansion and differentiation of T cells. Interfering with the signaling pathway downstream of CD28 could represent an alternative to biological blockade. The NF-κB cascade mediates the T cell-specific response to costimulation [46]. All B7 family coreceptors share a YXXM motif, allowing the inducible interaction with the p85 adaptor subunit of PI3K after phosphorylation by Lck. PI3K interacts with PDK1, which in turn recruits the IKK complex through PKCθ and CARMA1. PKCθ phosphorylates CARMA1, which leads to the assembly of the trimolecular complex CARMA1-BCL10-MALT1 that activates NF-κB.

PKCθ represents a potential T cell-specific therapeutic target. PKCθ-deficient mice are protected against T cell-dependent arthritis [47]. Pharmacological agents directed against components of the CARMA1-BCL10-MALT1 signalosome could interfere with both T and B cell function given the additional role of this complex in BCR signal transduction. Downstream of this molecular platform, the same core components of the NF-κB signaling pathway control T and B lymphocyte activation and the expression of most inflammatory cytokines, chemokines and proteolytic enzymes in non-lymphoid cells such as synovial fibroblasts, macrophages and osteoclasts. Thus, therapeutic strategies targeting the NF-κB pathway can potentially control both the innate and adaptive immune responses in RA.

Representative critical signaling pathways in RA

Existing anti-cytokines therapies block the interaction of a single ligand with its specific receptor. A significant percentage of RA patients do not respond to such therapies. Moreover, disease control or remission usually requires the concomitant use of broader immunosuppressive drugs such as methotrexate and/or steroids, exposing patients to significant side effects. This is not surprising given the redundancy of pro-inflammatory factors in the rheumatoid joint. However, these ligands rely on restricted set of core signaling pathways. Mapping the hierarchy of these pathways can identify specific targets that control the production of the most harmful set of inflammatory proteins and minimize potential toxicity.

NF-κB

NF-κB affects multiple aspects of the inflammatory and destructive processes in RA. In the RA synovium, activated NF-κB is detected in both macrophages and FLS, and induces the transcription of cytokines like IL-1, TNF and IL-6 [48–50]. NF-κB also plays an important role in leukocyte adhesion and transmigration by controlling the transcription of chemokines and cell adhesion molecules. In animal models of RA, NF-κB activation in the joint heralds the appearance of clinical disease. The expression of MMPs, but not their tissue inhibitor, is under the control of NF-κB, as well as the differentiation of T and B cells. Bone destruction is enhanced by RANK activation (receptor activator of NF-κB) in osteoclasts.

NF-κB is a family of transcription factors that increase the mRNA levels of its target genes by binding to specific DNA sequences called κB or Rel sites [51]. These DNA motifs are usually found in the promoter, a regulatory region of the gene close to the transcription start site. The five NF-κB members are classified in two subfamilies called NF-κB and Rel. The C-terminal part of NF-κB p105 and p100 proteins contain IκB-like inhibitory domains characterized by multiple copies of the ankyrin repeat (ANK). These precursors are processed by proteasome-mediated partial proteolysis to their mature DNA-binding forms p50 and p52, respectively. c-Rel, RelA and RelB are not processed. NF-κB transcription complexes are homo- or heterodimers of any of the subunits p50, p52, c-Rel, RelA (p65) and RelB.

The primary regulation of the NF-κB pathway is through the association of NF-κB complexes with IκB inhibitors that comprise IκBα, IκBβ, and IκBε. IκB bind to the Rel domain of NF-κB dimers and block the nuclear localization function of Rel. These inactive IκB-NF-κB multimeric complexes are thus maintained within the cytoplasm. The two main pathways leading to NF-κB activation are called the classical and alternative pathways [52]. Both are activated by an IκB kinase complex consisting of catalytic subunits (IKKα and/or IKKβ, also called IKK1 and IKK2). In addition, the classical IKK complex comprises the scaffold protein NF-κB essential modulator (NEMO) or IKKγ.

Innate immunity and inflammation relies strongly on the classical pathway [49]. Adaptors such as TRAF2 (downstream of TNFRI) and TRAF6 (downstream of IL-1R/TLR4) link to MAP/ERK kinase kinase 3 (MEKK3) and TAK1. This in turn activates IKKs, which then phosphorylate IκB inhibitors, targeting them for ubiquitination and degradation by the 26S proteasome. The main latent classical complex is a p50-RelA-IκBα trimer. The p50-RelA dimer drives most of the κB transcriptional activity.

Other receptors preferentially induce the alternative pathway, such as BAFF-R that controls B cell survival and RANK that increases osteoclastogenesis and bone destruction [52]. However, although IKKα is required for RANK ligand-induced osteoclast formation *in vitro*, it was dispensable *in vivo* in a mouse model [53]. The alternative IKK signalosome consists of IKKα subunits only, which are activated by NIK (NF-κB inducing kinase). The latent complex comprises RelB and p100. The slow degradation of p100 into the active p52/RelB dimers results in a delayed kinetic of onset (hours *versus* minutes for the classical pathway).

Different classes of compounds can inhibit the NF-κB pathway including peptides, oligonucleotides, microbial and viral proteins, small molecules or engineered dominant negative or constitutively active polypeptides [54]. Some inhibit NF-κB globally, while others target specific steps of the classical or the alternative pathway. Several immunosuppressive drugs are also able to modulate NF-κB activity. Glucocorticoids inhibit DNA binding of Rel or IKK activity depending on the cellular context. Cyclosporine interferes with the proteasome to decrease IκB degradation. FK506 blocks c-Rel nuclear translocation but not p50/RelA. Many other drugs show anti-NF-κB activity *in vitro* and/or *in vivo* sometimes at non-physiological doses like aspirin, suggesting an off-target effect or cross-talk between the drug's primary target and the NF-κB pathway. The development of more selective inhibitors will have the advantage to minimize NF-κB-independent toxicity.

IKK2 has received much of the focus in RA, in part because this kinase regulates cytokine production in many cell types including cultured synoviocytes [55]. Gene therapy approaches with dominant-negative forms of IKK2 are effective in animal models but they require local intra-articular therapy [56]. Systemic administration of several selective IKK2 inhibitors diminishes both inflammation and articular destruction in several rat or murine arthritis models [57]. No potent inhibitors of IKK1 have been described despite its unique role in the alternative NF-κB pathway, which is activated by surface receptors involved in RA such as RANK and BAFF-R. However, IKK1 might also contribute to the resolution of inflammatory responses by accelerating the removal of the NF-κB subunits RelA and c-Rel from gene promoters [58].

Downstream of IKK, ubiquitination or proteasome inhibitors prevent nuclear entry of NF-κB by blocking IκB degradation. The small molecule Ro196-9920 is an inhibitor of IκBα ubiquitination and was effective when administered orally in the mouse air pouch model [59]. The proteasome inhibitor PS-341 (now called bortezomib) reduces inflammation and bone and cartilage damage in rat streptococcal cell wall-induced arthritis [60]. This agent became the first proteasome inhibitor in clinical use for the treatment of multiple myeloma [61]. It is not clear whether NF-κB inhibition can explain all of its pro-apoptotic and anti-myeloma effects. The proteasome controls the stability of many proteins that regulate progression through the cell cycle and apoptosis, such as cyclins, cyclin-dependent kinases and tumor suppressors. These additional pro-apoptotic activities of proteasome inhibitors could be favorable in the treatment of RA.

Pathological resistance to apoptosis is thought to contribute to the proliferative phenotype of FLS. Inhibition of proteasome function in FLS increases spontaneous and TNF-induced apoptosis [62]. Despite the broad range of proteins that are processed by the proteasome, the toxicity of bortezomib in a Phase III clinical trial was restricted to dose-dependent and mostly reversible sensory neuropathies and thrombocytopenia [61]. It is not known if the dosing regimen currently used in cancer studies also produces anti-inflammatory effects.

Other methods for stabilizing IκB have been evaluated extensively in cell culture. IκB "super-repressors" are essentially IκB mutants engineered to lack the sites for phosphorylation, blocking their ubiquitination and degradation. A membrane permeable version of an IκB repressor could suppress local cytokine production and leukocytes influx during inflammation *in vivo* [63]. However, the bioavailability of this class of protein compounds limits their clinical use.

Downstream of IκB, some NF-κB inhibitors block its nuclear translocation or its transcriptional activity. DHMEQ, a fungal epoxyquinoid, has been reported to be a specific inhibitor of NF-κB nuclear translocation and was effective in collagen-induced arthritis [64]. Decoy oligonucleotides that contain an NF-κB-binding site showed efficacy in rat models of arthritis [65, 66]. However, their poor bioavailability limits their therapeutic potential.

In summary, NF-κB inhibitors could reduce several pathogenic processes in RA, including the resistance of FLS to apoptosis, and the increased inflammatory and autoimmune responses. On the other hand, NF-κB transcription factors also control the expression of genes that play key roles in cell survival and response to stress. These functions raise safety issues related to the systemic inhibition of this pathway. For instance, disrupting the function of IKK2 or Rela causes embryonic lethality in mice by increasing TNF induced hepatocyte apoptosis [67]. However, the toxicity of IKK2 blockade in the adult might be less than embryonic effects. In the adult liver, IKK1 is able to maintain sufficient levels of NF-κB activity to resist the stress induced by LPS or soluble TNF administration [68]. AS602868, a small molecule inhibitor of IKK2, has no effect on transaminases levels and protects mice from ischemia-reperfusion liver injury without sensitizing them toward TNF-induced apoptosis [69]. Intriguingly, adult IKK2 knockout hepatocytes remain sensitive to apoptosis induced by activated T cells that express the membrane-bound rather than the soluble form of TNF [68].

Because IKK2 and other members of the NF-κB pathway drive many aspects of adaptive and innate immune responses, the benefit provided by inhibition of NF-κB could be difficult to dissociate from the risk of infectious complications. One goal could be to modulate, rather than ablate, NF-κB. Selective targeting of NF-κB heterodimers or homodimers could restrict the effects to a subset of genes involved in autoinflammation. Alternatively, allowing a basal level of NF-κB activity to persist might be safer than blocking it.

MAPKs

MAPKs are a family of serine/threonine protein kinases that control an array of cellular responses to external stress signals. In RA, they regulate both cytokine production and cytokine action [70]. MAPK cascades are organized into three individual subfamilies: ERK, JNK and p38. The hierarchy of signaling events within each family is based on a similar linear scheme: MAPK kinases (MAPKK or MAP2K) activate p38, JNK or ERK, while MAPKK are themselves activated by MAPKK kinases (MAPKKK or MAP3K). However, complex parallel and crossover signaling occurs. For instance, the MAP3K MEKK3 can activate both MKK3 (a p38-specific MAP2K) and MKK4 (a JNK MAP2K). Thus, many extracellular signals activate more than one MAPK. For instance, cytokine receptors and growth factor receptors lead mainly to ERK activation, whereas TLRs, proinflammatory cytokines, and osmotic shock activate p38. Ultraviolet light, protein synthesis inhibitors, and cytokines such as IL-1 and TNF stimulate the JNK pathway.

All MAPK regulate the expression of several genes that participate in synovial inflammation, including TNF and IL-1, but also MMPs that are involved in articular destruction. JNK works mainly by inducing the phosphorylation of c-Jun, which stimulates the transcription of genes controlled by the transcription factor AP-1. p38 acts at several levels including increasing the transcription, stability or translation of the mRNA, depending on the cell type and the stimulus. The ERKs 1 and 2 are widely expressed and regulate cellular proliferation and differentiation. Thus, inhibition of the ERK pathway is considered to be mainly relevant to the treatment of malignancies.

p38 MAPK pathway

Of the four known p38 family members, p38α is the main isoform expressed in macrophages and synoviocytes, but p38δ is also expressed at sites of invasion into the extracellular matrix [70]. The activated, phosphorylated form of p38 is found in the intimal lining where most of the cytokines and proteases are produced. Phosphorylated MKK3 and MKK6, the two major upstream regulators of p38, are also localized to the intimal lining and, to a lesser extent, in perivascular lymphoid aggregates [71]. p38, MKK3 and MKK6 regulate the expression of many important genes in synovial inflammation, including TNF, IL-1, and MMPs as well as the response of FLS to these inflammatory cytokines [72]. p38 inhibitors decrease both the inflammation and the erosive, destructive process in several preclinical models of arthritis [73]. However, preclinical and clinical toxicity have interfered with their development. Only a few publications detail the results of p38 inhibitors in clinical trials, which were often limited because of hepatotoxicity and gastrointestinal side effects.

The p38α/β selective inhibitor RWJ-67657 significantly suppressed the fever response and partially blocked the increase in serum cytokines such as TNF after LPS challenge in humans [74]. This compound also blocks *in vitro* cytokine production by FLS [75]. Some results from clinical trials of p38 inhibitors in RA are available in the form of abstracts or press releases. VX-745 was tested in a placebo-controlled study involving 44 patients. Of the patients in the treatment arm, 43% achieved an American College of Rheumatology (ACR) 20 response compared with 17% for placebo. Exposure to the drug was limited by hepatotoxicity, which precluded the use of methotrexate. Another compound, VX-702, also provided a modest dose dependent increase in ACR 20 responders in a prospective study involving 278 patients during 12 weeks of treatment. The ACR 20 response rate was 44% compared with 31% for placebo. Hepatotoxicity was not reported as significant. Another study with the p38 inhibitor SCIO 469 also showed limited efficacy [76]. Perhaps more striking was the fact that acute phase reactants like C-reactive protein rapidly decreased in the SCIO 469-treated patients, but later returned to pre-treatment levels. A similar phenomenon was observed with BIRB 796 in inflammatory bowel disease. It is not clear why some components of the inflammatory response, such as acute phase reactants, appear to escape from p38 inhibition.

Among the dose-limiting side effects that contributed to the limited efficacy of p38 inhibitors, liver tissues could be targeted since structurally distinct compounds displayed hepatoxicity. However, this is not certain and greater selectivity for p38 over other kinases and p38 isoforms might improve the tolerability. Allosteric inhibitors might also help reduce off target toxicities. Another alternative would be to target upstream regulators or downstream substrates of p38. MKK3 deficiency suppresses inflammation and articular cytokine production in a passive murine arthritis model but does not affect IL-6 release after LPS injection. Thus, targeting this upstream kinase could protect from synovial inflammation, while leaving host defense responses intact [77]. MAPKAP kinase 2 (MK2) is a substrate of p38 involved in the post-transcriptional regulation of TNF biosynthesis. Deletion of the MK2 gene protects mice from collagen-induced arthritis [78]. However, this strategy decreases cytokine release after LPS challenge *in vivo* and might lower the resistance to microbial infection.

JNK pathway

Both JNK1 and JNK2 are phosphorylated in RA but not osteoarthritis synovium [79]. JNK3 expression is restricted to neurological tissue. Activated forms of MKK4 and MKK7, the upstream activators of JNK, are both expressed in rheumatoid synovium with its major substrate c-Jun [80]. JNK controls local inflammation in the joint by regulating cytokine and MMPs expression. Studies in JNK-deficient mice indicate that this MAPK also regulates T cell differentiation into the T helper 1 subset, which has been incriminated in the pathology of RA [81]. Thus, JNK inhibition could diminish adaptive autoimmune responses in RA in addition to blocking MMP production by FLS.

Peptide-based approaches that can target JNK signaling have been reported. Specific fragments of JNK-interacting protein 1 (JIP1), a scaffold protein, inhibit JNK activity in various cell types. Short JIP1-derived peptides are selective and appear to inhibit only JNK and its upstream activators, MKK4 and MKK7 [82]. Systemic administration of a cell-penetrating and protease-resistant peptide inhibitor of JIP1 can prevent neuronal degeneration in different *in vivo* models [82].

SP600125 is a reversible ATP competitive inhibitor for all three JNK isoforms. Administration of this compound during rat adjuvant arthritis had a modest anti-inflammatory effect. In contrast, radiographic analysis showed a dramatic decrease in bone and cartilage damage [83]. This was likely due to suppression of effector mechanisms such as MMP production in the inflamed joints because the treatment was started after initiation of T cell-dependent autoimmune responses. Data from mice deficient in either JNK1 or JNK2 suggests that both JNK isoforms need to be inhibited. Indeed, JNK2 deficiency offered very little protection in passive collagen arthritis, indicating that JNK1 can compensate [84]. Conversely, JNK1 knockout does not protect from synovial inflammation in a TNF transgenic model [85]. Another molecule, spleen tyrosine kinase (Syk), plays a key role in Fc receptor and ITAM signaling. Additional data suggest that it can also influence JNK function in cultured synoviocytes [86]. The Syk inhibitor R406/R788 has demonstrated efficacy in a Phase II study in RA patients concomitantly treated with methotrexate [87].

Long-term exposure to JNK inhibitors in chronic diseases could cause neurological toxicity. Mice lacking both JNK1 and JNK2 isoforms present late embryonic lethality and neural tube closure defects, while single deficiency in JNK1 elicit an age-dependent alteration of axons and dendrites associated with pronounced loss of microtubules [88, 89]. In cultured FLS, MKK7 but not MKK4 is required for JNK activation after cytokine stimulation [90]. This suggests that this upstream kinase might be an efficient and safer target than JNK in arthritis.

Conclusion

Aberrant regulation of several inflammatory pathways in RA provides a rationale for therapeutic intervention. Preclinical studies suggest several possible ways of interfering with these pathways including gene therapy approaches, small interfering RNA or inhibitory peptides. Small molecule compounds have several advantages in the clinic such as oral bioavailability and lower production costs. Inhibitors of NF-κB or MAPKs might require careful selection of the relevant kinases. Judicious dosing to modulate rather than block these pathways might also be required since they also perform important metabolic functions. Recently studies demonstrating efficacy for the Syk inhibitor R406/R788 in as well as beneficial effects of JAK1, JAK2, and JAK3 inhibitors suggest that targeted therapies focused on signal transduction molecules have great potential.

References

1   Burger D, Dayer JM, Palmer G, Gabay C (2006) Is IL-1 a good therapeutic target in the treatment of arthritis? *Best Pract Res Clin Rheumatol* 20: 879–896

2   Ji H, Pettit A, Ohmura K, Ortiz-Lopez A, Duchatelle V, Degott C, Gravallese E, Mathis D, Benoist C (2002) Critical roles for interleukin 1 and tumor necrosis factor α in antibody-induced arthritis. *J Exp Med* 196: 77–85

3   Horai R, Saijo S, Tanioka H, Nakae S, Sudo K, Okahara A, Ikuse T, Asano M, Iwakura Y (2000) Development of chronic inflammatory arthropathy resembling rheumatoid arthritis in interleukin 1 receptor antagonist-deficient mice. *J Exp Med* 191: 313–320

4   Bresnihan B, varo-Gracia JM, Cobby M, Doherty M, Domljan Z, Emery P, Nuki G,

Pavelka K, Rau R, Rozman B et al (1998) Treatment of rheumatoid arthritis with recombinant human interleukin-1 receptor antagonist. *Arthritis Rheum* 41: 2196–2204

5   Choe JY, Crain B, Wu SR, Corr M (2003) Interleukin 1 receptor dependence of serum transferred arthritis can be circumvented by toll-like receptor 4 signaling. *J Exp Med* 197: 537–542

6   Colotta F, Re F, Muzio M, Bertini R, Polentarutti N, Sironi M, Giri JG, Dower SK, Sims JE, Mantovani A (1993) Interleukin-1 type II receptor: A decoy target for IL-1 that is regulated by IL-4. *Science* 261: 472–475

7   Sims JE, Gayle MA, Slack JL, Alderson MR, Bird TA, Giri JG, Colotta F, Re F, Mantovani A, Shanebeck K et al (1993) Interleukin 1 signaling occurs exclusively *via* the type I receptor. *Proc Natl Acad Sci USA* 90: 6155–6159

8   Burns K, Janssens S, Brissoni B, Olivos N, Beyaert R, Tschopp J (2003) Inhibition of interleukin 1 receptor/Toll-like receptor signaling through the alternatively spliced, short form of MyD88 is due to its failure to recruit IRAK-4. *J Exp Med* 197: 263–268

9   Kobayashi K, Hernandez LD, Galan JE, Janeway CA Jr, Medzhitov R, Flavell RA (2002) IRAK-M is a negative regulator of Toll-like receptor signaling. *Cell* 110: 191–202

10 Bartfai T, Behrens MM, Gaidarova S, Pemberton J, Shivanyuk A, Rebek J Jr (2003) A low molecular weight mimic of the Toll/IL-1 receptor/resistance domain inhibits IL-1 receptor-mediated responses. *Proc Natl Acad Sci USA* 100: 7971–7976

11 Feldmann M, Maini RN (2001) Anti-TNF α therapy of rheumatoid arthritis: What have we learned? *Annu Rev Immunol* 19: 163–196

12 Aggarwal BB (2003) Signalling pathways of the TNF superfamily: A double-edged sword. *Nat Rev Immunol* 3: 745–756

13 Bai S, Liu H, Chen KH, Eksarko P, Perlman H, Moore TL, Pope RM (2004) NF-κB-regulated expression of cellular FLIP protects rheumatoid arthritis synovial fibroblasts from tumor necrosis factor α-mediated apoptosis. *Arthritis Rheum* 50: 3844–3855

14 De SE, Zazzeroni F, Papa S, Nguyen DU, Jin R, Jones J, Cong R, Franzoso G (2001) Induction of gadd45β by NF-κB downregulates pro-apoptotic JNK signalling. *Nature* 414: 308–313

15 Dunne A, O'Neill LA (2003) The interleukin-1 receptor/Toll-like receptor superfamily: Signal transduction during inflammation and host defense. *Sci STKE* 2003: re3

16 Radstake TR, Roelofs MF, Jenniskens YM, Oppers-Walgreen B, van Riel PL, Barrera P, Joosten LA, van den Berg WB (2004) Expression of toll-like receptors 2 and 4 in rheumatoid synovial tissue and regulation by proinflammatory cytokines interleukin-12 and interleukin-18 *via* interferon-γ. *Arthritis Rheum* 50: 3856–3865

17 Roelofs MF, Joosten LA, bdollahi-Roodsaz S, van Lieshout AW, Sprong T, van den Hoogen FH, van den Berg WB, Radstake TR (2005) The expression of toll-like receptors 3 and 7 in rheumatoid arthritis synovium is increased and costimulation of toll-like receptors 3, 4, and 7/8 results in synergistic cytokine production by dendritic cells. *Arthritis Rheum* 52: 2313–2322

18 Brentano F, Schorr O, Gay RE, Gay S, Kyburz D (2005) RNA released from necrotic

synovial fluid cells activates rheumatoid arthritis synovial fibroblasts *via* Toll-like receptor 3. *Arthritis Rheum* 52: 2656–2665

19 Sweeney SE, Mo L, Firestein GS (2007) Antiviral gene expression in rheumatoid arthritis: Role of IKKε and interferon regulatory factor 3. *Arthritis Rheum* 56: 743–752

20 Sweeney SE, Hammaker D, Boyle DL, Firestein GS (2005) Regulation of c-Jun phosphorylation by the IκB kinase-ε complex in fibroblast-like synoviocytes. *J Immunol* 174: 6424–6430

21 Kyburz D, Rethage J, Seibl R, Lauener R, Gay RE, Carson DA, Gay S (2003) Bacterial peptidoglycans but not CpG oligodeoxynucleotides activate synovial fibroblasts by toll-like receptor signaling. *Arthritis Rheum* 48: 642–650

22 Leadbetter EA, Rifkin IR, Hohlbaum AM, Beaudette BC, Shlomchik MJ, Marshak-Rothstein A (2002) Chromatin-IgG complexes activate B cells by dual engagement of IgM and Toll-like receptors. *Nature* 416: 603–607

23 Deng GM, Verdrengh M, Liu ZQ, Tarkowski A (2000) The major role of macrophages and their product tumor necrosis factor α in the induction of arthritis triggered by bacterial DNA containing CpG motifs. *Arthritis Rheum* 43: 2283–2289

24 Lee MS, Kim YJ (2007) Signaling pathways downstream of pattern-recognition receptors and their cross talk. *Annu Rev Biochem*

25 Mariathasan S, Monack DM (2007) Inflammasome adaptors and sensors: Intracellular regulators of infection and inflammation. *Nat Rev Immunol* 7: 31–40

26 Hoffman HM, Mueller JL, Broide DH, Wanderer AA, Kolodner RD (2001) Mutation of a new gene encoding a putative pyrin-like protein causes familial cold autoinflammatory syndrome and Muckle-Wells syndrome. *Nat Genet* 29: 301–305

27 Kanneganti TD, Lamkanfi M, Kim YG, Chen G, Park JH, Franchi L, Vandenabeele P, Nunez G (2007) Pannexin-1-mediated recognition of bacterial molecules activates the cryopyrin inflammasome independent of Toll-like receptor signaling. *Immunity* 26: 433–443

28 Miggin SM, Palsson-McDermott E, Dunne A, Jefferies C, Pinteaux E, Banahan K, Murphy C, Moynagh P, Yamamoto M, Akira S et al (2007) NF-κB activation by the Toll-IL-1 receptor domain protein MyD88 adapter-like is regulated by caspase-1. *Proc Natl Acad Sci USA* 104: 3372–3377

29 Eckmann L, Karin M (2005) NOD2 and Crohn's disease: Loss or gain of function? *Immunity* 22: 661–667

30 Saleh M, Mathison JC, Wolinski MK, Bensinger SJ, Fitzgerald P, Droin N, Ulevitch RJ, Green DR, Nicholson DW (2006) Enhanced bacterial clearance and sepsis resistance in caspase-12-deficient mice. *Nature* 440: 1064–1068

31 Razmara M, Srinivasula SM, Wang L, Poyet JL, Geddes BJ, DiStefano PS, Bertin J, Alnemri ES (2002) CARD-8 protein, a new CARD family member that regulates caspase-1 activation and apoptosis. *J Biol Chem* 277: 13952–13958

32 Chae JJ, Wood G, Masters SL, Richard K, Park G, Smith BJ, Kastner DL (2006) The B30.2 domain of pyrin, the familial Mediterranean fever protein, interacts directly with caspase-1 to modulate IL-1β production. *Proc Natl Acad Sci USA* 103: 9982–9987

33 Dinarello CA (2007) A signal for the caspase-1 inflammasome free of TLR. *Immunity* 26: 383–385

34 Hoffman HM, Rosengren S, Boyle DL, Cho JY, Nayar J, Mueller JL, Anderson JP, Wanderer AA, Firestein GS (2004) Prevention of cold-associated acute inflammation in familial cold autoinflammatory syndrome by interleukin-1 receptor antagonist. *Lancet* 364: 1779–1785

35 Aldea A, Campistol JM, Arostegui JI, Rius J, Maso M, Vives J, Yague J (2004) A severe autosomal-dominant periodic inflammatory disorder with renal AA amyloidosis and colchicine resistance associated to the MEFV H478Y variant in a Spanish kindred: An unusual familial Mediterranean fever phenotype or another MEFV-associated periodic inflammatory disorder? *Am J Med Genet A* 124: 67–73

36 Rosengren S, Mueller JL, Anderson JP, Niehaus BL, Misaghi A, Anderson S, Boyle DL, Hoffman HM (2007) Monocytes from familial cold autoinflammatory syndrome patients are activated by mild hypothermia. *J Allergy Clin Immunol* 119: 991–996

37 Kremer JM, Westhovens R, Leon M, Di Giorgio E, Alten R, Steinfeld S, Russell A, Dougados M, Emery P, Nuamah IF et al (2003) Treatment of rheumatoid arthritis by selective inhibition of T-cell activation with fusion protein CTLA4Ig. *N Engl J Med* 349: 1907–1915

38 Sakaguchi N, Takahashi T, Hata H, Nomura T, Tagami T, Yamazaki S, Sakihama T, Matsutani T, Negishi I, Nakatsuru S et al (2003) Altered thymic T-cell selection due to a mutation of the ZAP-70 gene causes autoimmune arthritis in mice. *Nature* 426: 454–460

39 Nel AE (2002) T-cell activation through the antigen receptor. Part 1: Signaling components, signaling pathways, and signal integration at the T-cell antigen receptor synapse. *J Allergy Clin Immunol* 109: 758–770

40 Viola A, Schroeder S, Sakakibara Y, Lanzavecchia A (1999) T lymphocyte costimulation mediated by reorganization of membrane microdomains. *Science* 283: 680–682

41 Bottini N, Musumeci L, Alonso A, Rahmouni S, Nika K, Rostamkhani M, MacMurray J, Meloni GF, Lucarelli P, Pellecchia M et al (2004) A functional variant of lymphoid tyrosine phosphatase is associated with type I diabetes. *Nat Genet* 36: 337–338

42 Vang T, Congia M, Macis MD, Musumeci L, Orru V, Zavattari P, Nika K, Tautz L, Tasken K, Cucca F et al (2005) Autoimmune-associated lymphoid tyrosine phosphatase is a gain-of-function variant. *Nat Genet* 37: 1317–1319

43 Morgan ME, Flierman R, van Duivenvoorde LM, Witteveen HJ, van EW, van Laar JM, de Vries RR, Toes RE (2005) Effective treatment of collagen-induced arthritis by adoptive transfer of CD25+ regulatory T cells. *Arthritis Rheum* 52: 2212–2221

44 Beyersdorf N, Hanke T, Kerkau T, Hunig T (2006) CD28 superagonists put a break on autoimmunity by preferentially activating CD4+CD25+ regulatory T cells. *Autoimmun Rev* 5: 40–45

45 Suntharalingam G, Perry MR, Ward S, Brett SJ, Castello-Cortes A, Brunner MD, Panoskaltsis N (2006) Cytokine storm in a phase 1 trial of the anti-CD28 monoclonal antibody TGN1412. *N Engl J Med* 355: 1018–1028

46   Weil R, Israel A (2006) Deciphering the pathway from the TCR to NF-κB. *Cell Death Differ* 13: 826–833

47   Healy AM, Izmailova E, Fitzgerald M, Walker R, Hattersley M, Silva M, Siebert E, Terkelsen J, Picarella D, Pickard MD et al (2006) PKC-theta-deficient mice are protected from Th1-dependent antigen-induced arthritis. *J Immunol* 177: 1886–1893

48   Han Z, Boyle DL, Manning AM, Firestein GS (1998) AP-1 and NF-κB regulation in rheumatoid arthritis and murine collagen-induced arthritis. *Autoimmunity* 28: 197–208

49   Handel ML, McMorrow LB, Gravallese EM (1995) Nuclear factor-κB in rheumatoid synovium. Localization of p50 and p65. *Arthritis Rheum* 38: 1762–1770

50   Miyazawa K, Mori A, Yamamoto K, Okudaira H (1998) Constitutive transcription of the human interleukin-6 gene by rheumatoid synoviocytes: Spontaneous activation of NF-κB and CBF1. *Am J Pathol* 152: 793–803

51   Karin M, Yamamoto Y, Wang QM (2004) The IKK NF-κB system: A treasure trove for drug development. *Nat Rev Drug Discov* 3: 17–26

52   Bonizzi G, Karin M (2004) The two NF-κB activation pathways and their role in innate and adaptive immunity. *Trends Immunol* 25: 280–288

53   Ruocco MG, Maeda S, Park JM, Lawrence T, Hsu LC, Cao Y, Schett G, Wagner EF, Karin M (2005) IκB kinase (IKK)β, but not IKKα, is a critical mediator of osteoclast survival and is required for inflammation-induced bone loss. *J Exp Med* 201: 1677–1687

54   Gilmore TD, Herscovitch M (2006) Inhibitors of NF-κB signaling: 785 and counting. *Oncogene* 25: 6887–6899

55   Andreakos E, Smith C, Kiriakidis S, Monaco C, de MR, Brennan FM, Paleolog E, Feldmann M, Foxwell BM (2003) Heterogeneous requirement of IκB kinase 2 for inflammatory cytokine and matrix metalloproteinase production in rheumatoid arthritis: Implications for therapy. *Arthritis Rheum* 48: 1901–1912

56   Tak PP, Gerlag DM, Aupperle KR, van de Geest DA, Overbeek M, Bennett BL, Boyle DL, Manning AM, Firestein GS (2001) Inhibitor of nuclear factor κB kinase β is a key regulator of synovial inflammation. *Arthritis Rheum* 44: 1897–1907

57   McIntyre KW, Shuster DJ, Gillooly KM, Dambach DM, Pattoli MA, Lu P, Zhou XD, Qiu Y, Zusi FC, Burke JR (2003) A highly selective inhibitor of IκB kinase, BMS-345541, blocks both joint inflammation and destruction in collagen-induced arthritis in mice. *Arthritis Rheum* 48: 2652–2659

58   Lawrence T, Bebien M, Liu GY, Nizet V, Karin M (2005) IKKα limits macrophage NF-κB activation and contributes to the resolution of inflammation. *Nature* 434: 1138–1143

59   Swinney DC, Xu YZ, Scarafia LE, Lee I, Mak AY, Gan QF, Ramesha CS, Mulkins MA, Dunn J, So OY et al (2002) A small molecule ubiquitination inhibitor blocks NF-κB-dependent cytokine expression in cells and rats. *J Biol Chem* 277: 23573–23581

60   Palombella VJ, Conner EM, Fuseler JW, Destree A, Davis JM, Laroux FS, Wolf RE, Huang J, Brand S, Elliott PJ et al (1998) Role of the proteasome and NF-κB in streptococcal cell wall-induced polyarthritis. *Proc Natl Acad Sci USA* 95: 15671–15676

61 Richardson PG, Sonneveld P, Schuster MW, Irwin D, Stadtmauer EA, Facon T, Harousseau JL, Ben-Yehuda D, Lonial S, Goldschmidt H et al (2005) Bortezomib or high-dose dexamethasone for relapsed multiple myeloma. *N Engl J Med* 352: 2487–2498

62 Kawakami A, Nakashima T, Sakai H, Hida A, Urayama S, Yamasaki S, Nakamura H, Ida H, Ichinose Y, Aoyagi T et al (1999) Regulation of synovial cell apoptosis by proteasome inhibitor. *Arthritis Rheum* 42: 2440–2448

63 Blackwell NM, Sembi P, Newson JS, Lawrence T, Gilroy DW, Kabouridis PS (2004) Reduced infiltration and increased apoptosis of leukocytes at sites of inflammation by systemic administration of a membrane-permeable IκBα repressor. *Arthritis Rheum* 50: 2675–2684

64 Wakamatsu K, Nanki T, Miyasaka N, Umezawa K, Kubota T (2005) Effect of a small molecule inhibitor of nuclear factor-κB nuclear translocation in a murine model of arthritis and cultured human synovial cells. *Arthritis Res Ther* 7: R1348–R1359

65 Tomita T, Takeuchi E, Tomita N, Morishita R, Kaneko M, Yamamoto K, Nakase T, Seki H, Kato K, Kaneda Y et al (1999) Suppressed severity of collagen-induced arthritis by *in vivo* transfection of nuclear factor κB decoy oligodeoxynucleotides as a gene therapy. *Arthritis Rheum* 42: 2532–2542

66 Miagkov AV, Kovalenko DV, Brown CE, Didsbury JR, Cogswell JP, Stimpson SA, Baldwin AS, Makarov SS (1998) NF-κB activation provides the potential link between inflammation and hyperplasia in the arthritic joint. *Proc Natl Acad Sci USA* 95: 13859–13864

67 Gerondakis S, Grumont R, Gugasyan R, Wong L, Isomura I, Ho W, Banerjee A (2006) Unravelling the complexities of the NF-κB signalling pathway using mouse knockout and transgenic models. *Oncogene* 25: 6781–6799

68 Maeda S, Chang L, Li ZW, Luo JL, Leffert H, Karin M (2003) IKKβ is required for prevention of apoptosis mediated by cell-bound but not by circulating TNFα. *Immunity* 19: 725–737

69 Luedde T, Assmus U, Wustefeld T, Meyer z, V, Roskams T, Schmidt-Supprian M, Rajewsky K, Brenner DA, Manns MP, Pasparakis M et al (2005) Deletion of IKK2 in hepatocytes does not sensitize these cells to TNF-induced apoptosis but protects from ischemia/reperfusion injury. *J Clin Invest* 115: 849–859

70 Sweeney SE, Firestein GS (2006) Mitogen activated protein kinase inhibitors: Where are we now and where are we going? *Ann Rheum Dis* 65 (Suppl 3): iii83–iii88

71 Chabaud-Riou M, Firestein GS (2004) Expression and activation of mitogen-activated protein kinase kinases-3 and -6 in rheumatoid arthritis. *Am J Pathol* 164: 177–184

72 Inoue T, Hammaker D, Boyle DL, Firestein GS (2005) Regulation of p38 MAPK by MAPK kinases 3 and 6 in fibroblast-like synoviocytes. *J Immunol* 174: 4301–4306

73 Badger AM, Griswold DE, Kapadia R, Blake S, Swift BA, Hoffman SJ, Stroup GB, Webb E, Rieman DJ, Gowen M et al (2000) Disease-modifying activity of SB 242235, a selective inhibitor of p38 mitogen-activated protein kinase, in rat adjuvant-induced arthritis. *Arthritis Rheum* 43: 175–183

74 Fijen JW, Zijlstra JG, de Boer P, Spanjersberg R, Tervaert JW, Van Der Werf TS,

Ligtenberg JJ, Tulleken JE (2001) Suppression of the clinical and cytokine response to endotoxin by RWJ-67657, a p38 mitogen-activated protein-kinase inhibitor, in healthy human volunteers. *Clin Exp Immunol* 124: 16–20

75 Westra J, Limburg PC, de Boer P, van Rijswijk MH (2004) Effects of RWJ 67657, a p38 mitogen activated protein kinase (MAPK) inhibitor, on the production of inflammatory mediators by rheumatoid synovial fibroblasts. *Ann Rheum Dis* 63: 1453–1459

76 Genovese MC, S. B. Cohen SB, D. Wofsy D, Weinblatt ME, Firestein GS, Brahn. Strand V, Baker DG, Tong SE (2008) Randomized, double-blind, placebo-controlled Phase 2 study of an oral p38α MAPK inhibitor, SCIO-469, in patients with active rheumatoid arthritis. *Arthritis Rheum* 58: S431

77 Inoue T, Boyle DL, Corr M, Hammaker D, Davis RJ, Flavell RA, Firestein GS (2006) Mitogen-activated protein kinase kinase 3 is a pivotal pathway regulating p38 activation in inflammatory arthritis. *Proc Natl Acad Sci USA* 103: 5484–5489

78 Hegen M, Gaestel M, Nickerson-Nutter CL, Lin LL, Telliez JB (2006) MAPKAP kinase 2-deficient mice are resistant to collagen-induced arthritis. *J Immunol* 177: 1913–1917

79 Han Z, Boyle DL, Aupperle KR, Bennett B, Manning AM, Firestein GS (1999) Jun N-terminal kinase in rheumatoid arthritis. *J Pharmacol Exp Ther* 291: 124–130

80 Sundarrajan M, Boyle DL, Chabaud-Riou M, Hammaker D, Firestein GS (2003) Expression of the MAPK kinases MKK-4 and MKK-7 in rheumatoid arthritis and their role as key regulators of JNK. *Arthritis Rheum* 48: 2450–2460

81 Dong C, Yang DD, Wysk M, Whitmarsh AJ, Davis RJ, Flavell RA (1998) Defective T cell differentiation in the absence of Jnk1. *Science* 282: 2092–2095

82 Borsello T, Clarke PG, Hirt L, Vercelli A, Repici M, Schorderet DF, Bogousslavsky J, Bonny C (2003) A peptide inhibitor of c-Jun N-terminal kinase protects against excito-toxicity and cerebral ischemia. *Nat Med* 9: 1180–1186

83 Han Z, Boyle DL, Chang L, Bennett B, Karin M, Yang L, Manning AM, Firestein GS (2001) c-Jun N-terminal kinase is required for metalloproteinase expression and joint destruction in inflammatory arthritis. *J Clin Invest* 108: 73–81

84 Han Z, Chang L, Yamanishi Y, Karin M, Firestein GS (2002) Joint damage and inflammation in c-Jun N-terminal kinase 2 knockout mice with passive murine collagen-induced arthritis. *Arthritis Rheum* 46: 818–823

85 Koller M, Hayer S, Redlich K, Ricci R, David JP, Steiner G, Smolen JS, Wagner EF, Schett G (2005) JNK1 is not essential for TNF-mediated joint disease. *Arthritis Res Ther* 7: R166–R173

86 Cha HS, Boyle DL, Inoue T, Schoot R, Tak PP, Pine P, Firestein GS. (2006) A novel spleen tyrosine kinase inhibitor blocks c-Jun N-terminal kinase-mediated gene expression in synoviocytes. *J Pharmacol Exp Ther*. 317: 571–8

87 Weinblatt ME, Kavanaugh A, Grossbard E, for the TASKI-1 Trial (2008) Rheumatoid arthritis (RA) with a Syk kinase inhibitor: A 12 week randomized placebo controlled study. *Arthritis Rheum* 58: S610

88 Chang L, Jones Y, Ellisman MH, Goldstein LS, Karin M (2003) JNK1 is required for

maintenance of neuronal microtubules and controls phosphorylation of microtubule-associated proteins. *Dev Cell* 4: 521–533

89 Kuan CY, Yang DD, Samanta Roy DR, Davis RJ, Rakic P, Flavell RA (1999) The Jnk1 and Jnk2 protein kinases are required for regional specific apoptosis during early brain development. *Neuron* 22: 667–676

90 Inoue T, Hammaker D, Boyle DL, Firestein GS (2006) Regulation of JNK by MKK-7 in fibroblast-like synoviocytes. *Arthritis Rheum* 54: 2127–2135

Targeting oncostatin M in the treatment of rheumatoid arthritis

Theresa C. Barnes and Robert J. Moots

Division of Inflammation, School of Clinical Sciences, University of Liverpool, University Hospital Aintree, Longmoor Lane, Liverpool L9 7AL, UK

Abstract

Oncostatin M (OSM) is a pleiotropic cytokine with potential utility as a treatment for inflammatory arthritis. This pro-inflammatory cytokine is increased in the rheumatoid but not in the osteoarthritic joint. Strategies to block the actions of OSM for use in inflammatory arthritis are being developed and these show significant promise in murine models of disease. Targeting OSM may have a beneficial effect by inhibiting some of the mechanisms of joint destruction in rheumatoid arthritis, which may limit long-term disability in patients. The challenge now is to convert this potential into firm compounds that can be tested in the clinic.

Introduction

Oncostatin M (OSM) is a 28-kDa cytokine with similarity to IL-6. In common with members of the IL-6 family of cytokines [IL-6, leukaemia inhibitory factor (LIF)] OSM signals through gp130 receptors. The type one receptor is composed of gp130 and LIFR beta and the type 2 receptor is composed of gp130 and OSMR beta. Common actions of LIF and OSM are signalled through the type 1 receptor, whereas OSM-specific actions are signalled through the type 2 receptor. Signalling through gp130 receptors activate JAK1, JAK 2 or Tyk2, which in turn activate STAT transcriptional activator proteins. In addition, OSM can signal through the MAPK signalling cascade (Fig. 1) [1]. OSM-specific effects are mediated through activation of specific STAT isoforms, which are selected, not by the specific JAK activation, but by interaction with tyrosine-based motifs within the OSM receptor. OSM is produced by activated T cells and macrophages and is also stored in the granules of neutrophils where it is released on degranulation [2].

OSM has pleiotropic effects including: stimulating the production of acute phase proteins in the liver, increasing the expression of adhesion molecules on endothelial cells, promoting IL-6 production by endothelial cells, inhibiting the growth and differentiation of some tumour cell lines, promoting the differentiation and proliferation of osteoblasts, inducing the formation of osteoclasts and regulating the

New Therapeutic Targets in Rheumatoid Arthritis, edited by Paul-Peter Tak
© 2009 Birkhäuser Verlag Basel/Switzerland

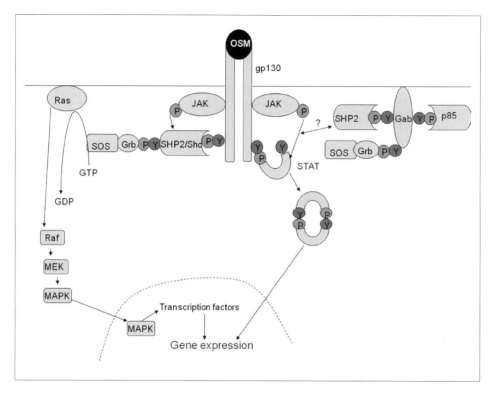

Figure 1
OSM binds to a gp130 receptor and signals through the JAK/STAT and MAPK pathways to alter gene expression (reproduced with permission from Heinrich et al (2003) Principles of interleukin (IL)-6-type cytokine signalling and its regulation. Biochem J *374: 1–20. © The Biochemical Society.*

expression of certain proteases including the matrix metalloproteinases (MMPs) and their naturally occurring inhibitors the tissue inhibitors of metalloproteinases (TIMPs) (Tab. 1).

The role of OSM in inflammation has been the subject of some debate. Some authors believe that the up-regulation of acute phase proteins and TIMPs reflects an anti-inflammatory role [3]. Others point to the effect on endothelial cells, osteoblasts and MMPs as evidence that OSM is, indeed, a predominantly inflammatory cytokine [4]. Further confusion has arisen because there appears to be variation in the responses to OSM between different species.

Therapeutic blockade of OSM has been explored in animal models of rheumatoid arthritis (RA), where it appears to ameliorate inflammation and, in particular, reduces tissue destruction. These observations suggest that this molecule might be

Table 1. OSM has pleiotropic actions on many different tissues. Outlined here are the main established actions.

Action of OSM	Tissue	Reference
Inhibit tumor growth	Melanoma cell lines	Malik et al 1989 [30]
Increased IL-6 production	Endothelial cells	Brown et al 1991 [31]
Induces plasminogen activator	Synovial fibroblasts	Hamilton et al 1991 [32]
Stimulates acute phase protein production	Liver	Richards et al 1992 [33]
Regulates TIMP expression	Synovial fibroblasts	Richards et al 1993 [34]
Increase in P-selectin expression	Endothelial cells	Yao et al 1996 [35]
Stimulates proteoglycan resorption	Cartilage	Wang Hui et al 1996 [9]
Inhibits proteoglycan synthesis	Cartilage	Wang Hui et al 1996 [9]
Synergise with IL-1 to increase MMP expression	Chondrocytes	Cawston et al 1998 [7]
Induces formation of osteoclasts	Bone	Richards et al 2000 [36]
Neutrophil adhesion and chemotaxis	Neutrophils	Kerfoot et al 2001 [37], Modur et al 1997 [4]
Differentiation and proliferation of osteoblasts	Bone	De Hooge et al 2002 [38], Palmqvist et al 2002 [39]
Cell proliferation	Endothelial cells and synovial fibroblasts	Fearon et al 2006 [19]
Synergise with IL-1 to increase VEGF production	Synovial fibroblast/ cartilage co-cultures	Fearon et al 2006 [19]
Increase endothelial cell tubule formation and migration	Endothelial cells	Fearon et al 2006 [19]

a relevant therapeutic target for RA and proof of concept studies are under way in humans to demonstrate whether this may be beneficial to patients with this disease.

Potential role of OSM in RA

Production of OSM in rheumatoid joints

Various studies have demonstrated that OSM is elevated in the rheumatoid synovium and synovial fluid. A variety of cells have been shown to produce OSM, including activated T cells and macrophages [5, 6] especially within the synovium [7].

Neutrophils, however, are the most abundant inflammatory cell found in the rheumatoid joint. These cells are not only found in synovial effusions, but also in the synovial pannus of patients with RA, particularly at the site of bony erosions, where they appear to have degranulated [8]. We have shown that neutrophils are capable of expressing and releasing significant amounts of OSM in response to stimulation with granulocyte/macrophage colony-stimulating factor (GM-CSF) [2]. Furthermore, we have found that neutrophils isolated from the synovial fluid of patients with RA are unresponsive, in terms of expressing and secreting OSM, to GM-CSF *in vitro*. This observation was in contrast to neutrophils isolated from the peripheral blood of the same patients, which expressed and secreted OSM on stimulation with GM-CSF as normal. This may represent *in vivo* stimulation of neutrophils to express and release OSM rendering them insensitive to further *in vitro* stimulation. Given the vast abundance of neutrophils in RA synovial fluid, these cells appear to be the major contributor of OSM in this compartment [2].

In vitro models of bone erosion

The inflammatory response in RA ultimately leads to cartilage and bone destruction, represented macroscopically by erosions. Such erosion of the joint tissues, when established, is irreversible and the panacea of treatment is to prevent this damage from occurring. OSM may play a potential role in mediating joint damage. It has been shown to increase the release of proteoglycans from pig cartilage explants *in vitro* coupled with a decrease in proteoglycan synthesis [9], indicating cartilage catabolism, which may result in cartilage destruction.

OSM is found in the synovial fluid of patients with RA but not in osteoarthritis [10]. In the synovial fluid of patients with RA, the concentration of OSM correlates with synovial fluid neutrophil counts [10]. Concentrations of OSM in rheumatoid synovial fluid at first sight appear to be too low to reproduce the *in vitro* effects on cartilage destruction. However, given that OSM and IL-1 act synergistically on cartilage destruction and proteoglycan synthesis, these levels may well be of pathological relevance. OSM expression is also increased in the synovial tissue of patients with RA [7]. In RA synovial fluid, concentrations of OSM correlate with synovial fluid keratin sulphate and pyridinoline levels, which indicate cartilage catabolism [11].

MMPs, OSM and RA

MMPs are protease enzymes that can degrade all elements of the extracellular matrix. Their activity is strictly controlled at several points, including synthesis, secretion, processing of the inactive proform and binding to specific inhibitors (TIMPs). MMPs have been found in increased concentrations in the serum, synovial fluid and synovial tissue of RA patients, where levels correlate with disease activity and structural damage [12–16]. The production of MMPs with potent collagenase activity (MMP-1 and MMP-13) is increased in RA synovial fibroblasts (RASFs) by the cytokines IL-1 and TNF-α [17]. OSM and retinoic acid both result in an increase in bovine and porcine cartilage degradation, as assessed by collagen and proteoglycan release. Together they act synergistically to produce cartilage degradation and increase expression of MMPs. Neither agent alone or in combination was able to lead to human cartilage degradation *in vitro* [18], reinforcing the importance of considering species differences in response to this cytokine. OSM and IL-1 synergistically increase collagen and proteoglycan release from porcine, bovine and human cartilage in the presence of increased MMP activity. Although OSM alone increases TIMP activity, IL-1 and OSM in combination lead to a net decrease in TIMP activity. This combination of OSM with IL-1 is, so far, the only combination of cytokines shown to promote collagenolysis in human cartilage [7].

OSM and the rheumatoid joint microenvironment

The effects of OSM and IL-1 alone, and in combination on human dermal microvascular endothelial cells (HDMECs) and RASFs, have been studied *in vitro*. RASF and HDMEC proliferation increase in response to OSM and IL-1 alone with a synergistic effect when these cytokines are combined. OSM and IL-1, whether alone or in combination, increase the percentage of HDMECs and RASFs expressing ICAM-1, an important adhesion molecule for leucocytes enhancing migration into sites of inflammation.

Angiogenesis is an essential part of the development of inflammatory RA pannus. *In vitro*, OSM increases the formation of endothelial cell tubules and increases HDMEC migration, indicating that it provides an angiogenic stimulus. OSM and IL-1 together result in a modest but consistent increase in vascular endothelial growth factor (VEGF) in cartilage. The fact that VEGF is a potent angiogenic stimulus further supports the evidence that OSM may be a regulator of angiogenesis [19].

Co-culture of RASFs with chondrocytes dictates the response of each cell type to cytokines and leads to higher baseline MMP production compared to each cell type alone. In addition, co-culture results in higher levels of MMP production when stimulated with OSM and IL-1, than either cell type alone. When cartilage is incubated alone with OSM and IL-1 there appears to be a decrease in TIMP production. However, in co-cultures of RASF with cartilage, OSM and IL-1 result in a dramatic shift in the MMP:TIMP ratio in favour of the MMPs. Thus, the effects of the microenvironment, including the different cells types and expressed cytokines, need to be taken into account during *in vitro* experiments to be able to make a sensible prediction as to what is truly happening *in vivo* [19]. These experiments indicate that, although OSM can increase TIMP expression, within the context of the rheumatoid joint, TIMP is decreased – suggesting that the net effect of OSM is proinflammatory.

OSM in animal models of inflammatory arthritis

The injection of human recombinant OSM into synovial joints of goats has been found to produce clinical features of joint inflammation, with an influx of leucocytes into the synovial fluid [20]. In this model, there was a significant reduction in cartilage proteoglycan content, indicating proteoglycan release, coupled with a decrease in *ex vivo* proteoglycan production. Some of these features were abrogated by the co-administration of the IL-1 decoy receptor IL-1Ra, which blocks the action of IL-1. However, even in the presence of IL-1Ra there was significant joint inflammation and reduction of cartilage proteoglycan content in response to intra-articular OSM, indicating that not all of the effects of OSM could be mediated by an increase in IL-1 [20].

Hui et al. [21] overexpressed murine OSM in combination with either IL-1 or TNF in the joints of a murine host using adenoviral gene transfer. Each cytokine on its own induced moderate synovial hyperplasia and bone erosions. However, both combinations resulted in a significant increase in synovial hyperplasia, inflammation and bone damage. In addition, the combinations induced synergistic osteoclast formation and activation and an increase in the expression of RANK and RANKL in inflammatory cells, inflamed synovium and articular cartilage. RANKL is a member of the TNF superfamily. It signals through its receptor, RANK and is a key factor in the differentiation and activation of osteoclasts, the main cells responsible for bone resorption. Other studies from this group have shown that overexpression of these combinations of cytokines in mice led to a synergistic increase in joint inflammation with bone destruction and proteoglycan and collagen release [22]. In addition, there was a synergistic increase in MMP expression in the cartilage and synovium. While TIMP expression was increased by OSM alone, this effect was abrogated when OSM and IL-1 were co-expressed [23].

The ability of synovial fibroblasts from patients with RA to proliferate and form colonies in anchorage-independent conditions is thought to reflect their ability to contribute to aggressive pannus formation *in vivo*. Langdon et al. [24] overexpressed murine OSM in mouse synovial fibroblasts by adenoviral gene transfer and found that OSM significantly increased the ability of mouse synovial fibroblasts to form colonies in anchorage-independent conditions. Like other groups, they found that intra-articular administration of the murine OSM *via* an adenoviral gene transfer technique resulted in pronounced joint inflammation and cartilage erosion. They also noted that the effects outlasted the predicted effect of the adenoviral vector, indicating a sustained change in phenotype towards one of chronic inflammation [24].

OSM as candidate therapeutic target in inflammatory arthritis

Inhibition of OSM has been investigated in a number of murine models of inflammatory arthritis, with promising results. Collagen-induced arthritis (CIA) is a murine model of RA induced by immunisation of DBA/1 mice with bovine type 2 collagen. In pristane-induced arthritis (PIA) joint inflammation in mice is induced by the intraperitoneal administration of pristane. Although neither model mimics human RA perfectly, it is usual for new targets to be evaluated in these models to establish "proof of concept" before moving on to human studies. Mice with CIA and PIA have been treated with a neutralising antibody to OSM. Mice with CIA were treated with two doses of the anti-OSM antibody after clinical features of arthritis had developed. This protocol was effective at preventing further disease progression in these mice [25]. In the case of PIA, mice were dosed with intraperitoneal anti-OSM before the onset of clinical arthritis. This regime prevented the development of an inflammatory arthritis in these mice [25]. Therefore, neutralisation of OSM by antibody seems to be effective at both treating and preventing the development of arthritis in these two murine models of RA [25]. Other studies have shown contradicting results: when mice were administered human OSM it appeared to ameliorate symptoms in a mouse model of inflammatory arthritis [26]. However, it is important to remember that human OSM does not bind to the murine OSM receptor, it only binds to the murine LIF receptor [27]; therefore, the anti-inflammatory actions seen in this model should be interpreted with caution, as they only represent the results of signalling through the LIF receptor [27].

Clinical development of compounds to target OSM for RA are at very early stages. Scientists from GlaxoSmithKleine have reported the development of RNA aptamers, which are potent and selective antagonists of human OSM, preventing it binding to its gp130 receptor. In HepG2 reporter gene experiments these aptamers were shown to be capable of functionally antagonising OSM. These compounds could be used to probe the role of OSM in RA and other inflammatory conditions and might be developed as therapeutic agents in their own right [28].

Conclusions

OSM is a pleiotropic cytokine whose net actions in the rheumatoid joint appear to be pro-inflammatory. One of the most important actions of OSM is to act alone or synergistically with other proinflammatory cytokines (IL-1, TNF-α) to increase the ratio of MMPs:TIMPs. This results in catabolism of cartilage and bone, which is crucial for joint erosion and destruction. In addition, OSM appears to have an important role in angiogenesis. Angiogenesis within the rheumatoid pannus is not only essential to support the proliferation of the pannus but also promotes inflammatory cell infiltration of the joint in conjunction with endothelial cell activation. Neutrophils, macrophages and T cells once recruited to the rheumatoid joint act as a source of further OSM, potentially causing a positive feedback loop that may result in the chronic inflammation characteristic of RA.

The actions of OSM vary subtly, depending both on the species and the microenvironment. Like other cytokines, its actions are modulated by other cytokines that are inevitably co-expressed within the rheumatoid joint and its effects should be understood within the context of both species and microenvironment. The effects of OSM in RA and on inflamed rheumatoid joints suggest that this cytokine might be a suitable candidate for development as a therapeutic target for the future.

References

1   Li WQ, Dehnade F, Zafarullah M (2001) Oncostatin M-induced matrix metalloproteinase and tissue inhibitor of metalloproteinase-3 genes expression in chondrocytes requires janus kinase/STAT signalling pathway. *J Immunol* 166: 3491–3498

2   Cross A, Edwards SW, Bucknall R, Moots RJ (2004) Secretion of oncostatin M by neutrophils in rheumatoid arthritis. *Arthritis Rheum* 50: 1430–1436

3   Wahl AF, Wallace PM (2001) Oncostatin M in the anti-inflammatory response. *Ann Rheum Dis* 60: 75–80

4   Modur V, Feldhaus MJ, Weyrich AS, Jicha DL, Prescott SM, Zimmerman GA, McIntyre TM (1997) Oncostatin M is a proinflammatory mediator. *In vivo* effects correlate with endothelial cell expression of inflammatory cytokines and adhesion molecules. *J Clin Invest* 100: 158–168

5   Brown TJ, Lioubin MN, Marquardt H (1987) Purification and characterization of cytostatic lymphokines produced by activated human T lymphocytes: Synergistic antiproliferative activity of transforming growth factor β1, interferon γ and oncostatin M for human melanoma cells. *J Immunol* 139: 2977–2983

6   Zarling JM, Shoyab M, Marquardt H, Hanson MB, Lioubin N, Todard GJ (1986) Oncostatin M: A growth regulator produced by differentiated histiocyte lymphoma cells. *Proc Natl Acad Sci USA* 83: 9739–9743

7   Cawston TE, Curry VA, Summers CA, Clark IM, Riley GP, Life PF, Spaull JR, Goldring MB, Koshy PJT, Rowan AD, Shingleton WD (1998) The role of oncostatin M in animal and human connective tissue collagen turnover and its localisation within the rheumatoid joint. *Arthritis Rheum* 41: 1760–1771

8   Mohr W, Menninger H (1980) Polymorphonuclear granulocytes at the pannus-cartilage junction in rheumatoid arthritis. *Arthritis Rheum* 23: 1413–1414

9   Hui W, Bell M, Carroll G (1996) Oncostatin M (OSM) stimulates resorption and inhibits synthesis of proteoglycan in porcine articular cartilage explants. *Cytokine* 8: 495–500

10   Hui W, Bell M, Carroll G (1997) Detection of oncostatin M in synovial fluid from patients with rheumatoid arthritis. *Ann Rheum Dis* 56: 184–187

11   Manicourt DH, Poilvache P, van Egeren A, Devogelaer JP, Lenz ME, Thonar EJMA (2000) Synovial fluid levels of tumor necrosis factor α and oncostatin M correlate with levels of markers of the degradation of crosslinked collagen and cartilage aggrecan in rheumatoid arthritis but not in osteoarthritis. *Arthritis Rheum* 43: 281–288

12   Clark IM, Powell LK, Ramsay S, Hazleman BL, Cawston TE (1993) The measurement of collagenase, tissue inhibitor of metalloproteinases (TIMP), and collagenase-TIMP complex in synovial fluids from patients with osteoarthritis and rheumatoid arthritis. *Arthritis Rheum* 36: 372–379

13   Ishiguro N, Ito T, Obata K, Fujimoto N, Iwata H (1996) Determination of stromelysin-1, 72 and 92 kDa type IV collagenase, tissue inhibitor of metalloproteinase-1 (TIMP-1), and TIMP-2 in synovial fluid and serum from patients with rheumatoid arthritis. *J Rheumatol* 23: 1599–1604

14   Tetlow LC, Woolley DE (1998) Comparative immunolocalization studies of collagenase 1 and collagenase 3 production in the rheumatoid lesion, and by human chondrocytes and synoviocytes *in vitro*. *Br J Rheumatol* 37: 64–70

15   Yoshihara Y, Nakamura H, Obata K, Yamada H, Hayakawa T, Fujikawa K, Okada Y (2000) Matrix metalloproteinases and tissue inhibitors of metalloproteinases in synovial fluids from patients with rheumatoid arthritis or osteoarthritis. *Ann Rheum Dis* 59: 455–461

16   Cunnane G, Fitzgerald O, Beeton C, Cawston TE, Bresnihan B (2001) Early joint erosions and serum levels of matrix metalloproteinase 1, matrix metalloproteinase 3, and tissue inhibitor of metalloproteinases 1 in rheumatoid arthritis. *Arthritis Rheum* 44: 2263–2274

17   MacNaul KL, Chartrain N, Kark M, Tocci MJ, Hutchinson NI (1990) Discoordinate expression of stromelysin, collagenase, and tissue inhibitor of metalloproteinases-1 in rheumatoid human synovial fibroblasts. Synergistic effects of interleukin-1 and tumor necrosis factor-alpha on stromelysin expression. *J Biol Chem* 265: 17238–17345

18   Shingleton WD, Jones D, Xu X, Cawston TE, Rowan AD (2006) Retinoic acid and oncostatin M combine to promote cartilage degradation *via* matrix metalloproteinase-13 expression in bovine but not human chondrocytes. *Rheumatology* 45: 958–965

19   Fearon U, Mullan R, Markham T, Connolly M, Sullivan S, Poole AR, Fitzgerald O, Bresnihan B, Veale DJ (2006) Oncostatin M induces angiogenesis and cartilage degradation in rheumatoid arthritis synovial tissue and human cartilage cocultures. *Arthritis Rheum* 54: 3152–3162

20   Bell MC, Carroll GJ, Chapman HM, Mills JN, Hui Wang (1999) Oncostatin M induces leukocyte infiltration and cartilage proteoglycan degradation *in vivo* in goat joints. *Arthritis Rheum* 42: 2543–2551

21   Hui Wang, Cawston TE, Richards CD, Rowan AD (2005) A model of inflammatory arthritis highlights a role for oncostatin M in pro-inflammatory cytokine-induced bone destruction *via* RANK/RANKL. *Arthritis Res Ther* 7: R57–R64

22   Rowan AD, Hui Wang, Cawston TE, Richards CD (2003) Adenoviral gene transfer of interleukin-1 in combination with oncostatin M induces significant joint damage in a murine model. *Am J Pathol* 162: 1975–1984

23   Hui W, Rowan AD, Richards CD, Cawston TE (2003) Oncostatin M in combination with tumor necrosis factor alpha induces cartilage damage and matrix metalloproteinase expression *in vitro* and *in vivo*. *Arthritis Rheum* 48: 3404–3418

24   Langdon C, Kerr C, Hassen M, Hara T, Arsenault AL, Richards CD (2000) Murine oncostatin M stimulates mouse synovial fibroblasts *in vitro* and induces inflammation and destruction in mouse joints *in vivo*. *Am J Pathol* 157: 1187–1196

25   Plater-Zyberk C, Buckton J, Thompson S, Spaull J, Zanders E, Papworth J, Life PF (2001) Amelioration of arthritis in two murine models using antibodies to oncostatin M. *Arthritis Rheum* 44: 2697–2702

26   Wallace PM, MacMaster JF, Rouleau KA, Brown TJ, Loy JK, Donaldson KL, Wahl AF (1999) Regulation of inflammatory responses by oncostatin M. *J Immunol* 162: 5547–5555

27   Ichihara M, Hara T, Kim H, Murate T, Miyajima A (1997) Oncostatin M and leukemia inhibitory factor do not use the same functional receptor in mice. *Blood* 90: 165–173

28   Rhodes A, Deakin A, Spaull J, Coomber B, Aitken A, Life P, Rees S (2000) The generation and characterisation of antagonist RNA aptamers to human oncostatin M. *J Biol Chem* 275: 28555-28561

29   Heinrich PC, Behrmann I, Haan S, Hermanns HM, Muller-Newen G, Schaper F (2003) Principles of interleukin (IL)-6-type cytokine signalling and its regulation. *Biochem J* 374: 1–20

30   Malik N, Kallestad JC, Gunderson NL, Austin SD, Neubauer MG, Ochs V, Marquardt H, Zarling JM, Shoyab M, Wei CM et al (1989) Molecular cloning, sequence analysis and functional expression of a novel growth regulator oncostatin M. *Mol Cell Biol* 9: 2847–53

31   Brown TJ, Rowe JM, Liu JW, Shoyab M (1991) Regulation of IL-6 expression by oncostatin M. *J Immunol* 147: 2175–80

32   Hamilton JA, Leizer T, Piccoli DS, Royston KM, Butler DM, Croatto M (1991) Oncostatin M stimulates urokinase-type plasminogen activator activity in human synovial fibroblasts. *Biochem Biophys Res Commun* 180(2): 652–9

33   Richards CD, Brown TJ, Shoyab M, Baumann H, Gauldie J (1992) Recombinant oncostatin M stimulates the production of acute phase proteins in HepG2 cells and rat primary hepatocytes *in vitro*. *J Immunol* 148: 1731–6

34   Richards CD, Shoyab M, Brown TJ, Gauldie J (1993) Selective regulation of metalloproteinase inhibitor (TIMP-1) by oncostatin M in fibroblasts in culture. *J Immunol* 150: 5596–603

35   Yao L, Pan J, Setiadi H, Patel KD, McEver RP (1996) Interleukin 4 or oncostatin M induces a prolonged increase in P-selectin mRNA and protein in human endothelial cells. *J Exp Med* 184: 81–92

36   Richards CD, Langdon C, Deschamps P, Pennica D, Shaughnessy SG (2000) Stimulation

of osteoclast differentiation *in vitro* by mouse oncostatin M, leukaemia inhibitory factor, cardiotrophin-1 and interleukin 6: synergy with dexamethasone. *Cytokine* 12: 613–21

37 Kerfoot SM, Raharjo E, Ho M, Kaur J, Serirom S, McCafferty DM, Burns AR, Patel KD, Kubes P (2001) Exclusive neutrophil recruitment with oncostatin M in a human system. *Am J Pathol* 159: 1531–9

38 De Hooge AS, van de Loo FA, Bennink MB, de Jong DS, Arntz OJ, Lubberts E, Richards CD, van den Berg WB (2002) Adenoviral transfer of murine oncostatin M elicits periosteal bone apposition in knee joints of mice, despite synovial inflammation and up-regulated expression of interleukin-6 and receptor activator of nuclear factor-kappa B ligand. *Am J Pathol* 160: 1733–43

39 Palmqvist P, Persson E, Conaway HH, Lerner UH (2002) IL-6, leukemia inhibitory factor, and oncostatin M stimulate bone resorption and regulate the expression of receptor activator of NF-kappa B ligand, osteoprotegerin, and receptor activator of NF-kappa B in mouse calvariae. *J Immunol* 169: 3353–62

Targeting the epigenetic modifications of synovial cells

Lars C. Huber, Astrid Jüngel and Steffen Gay

Center of Experimental Rheumatology, Department of Rheumatology, University Hospital of Zürich, Zürich Center of Integrative Human Physiology (ZIHP), Gloriastrasse 25, 8091 Zürich, Switzerland

Abstract

Rheumatoid arthritis (RA) is a systemic inflammatory disease that mainly affects the synovial tissues of joints. As in other autoimmune-related disorders, neither the etiology nor the pathogenesis of RA has as yet been completely unraveled. It is generally accepted, however, that autoimmune disorders develop through a combination of the individual genetic susceptibility, environmental factors, and dysregulated immune responses. Genetic predisposition has been described in RA, in particular as "shared epitope", a distinct sequence of amino acids within the antigen-presenting peptide groove of the major histocompatibility complex. Imbalanced immunity is reflected by the production of autoantibodies and the accumulation of reactive helper T cells within the rheumatoid synovium. In addition, environmental factors have been postulated as disease-modulating agents, including smoking, nutrition and infectious agents. So far, these factors have been studied almost exclusively as separate agents. However, gene transcription might be affected by ageing and environmental effects (such as nutrition and infections) – without changes in the nucleotide sequence of the underlying DNA. These patterns of alterations in the gene expression profiles are called "epigenetics". The term epigenetics is used to refer to molecular processes that regulate gene expression patterns but without changing the DNA nucleotide sequence. These epigenetic changes comprise the post-synthetic methylation of DNA and post-transcriptional modifications of histones, including methylation, phosphorylation, ubiquitination, sumoylation, biotinlyation and, most importantly, deacetylation and acetylation. With respect to the complex pathogenesis of rheumatic diseases, the "epigenome" is an emerging concept that integrates different etiologies and, thus, offers the opportunity for novel therapeutic strategies. Based on the fact that current therapies have not resulted in an ACR70 above 60% and have never been targeting the activated synovial fibroblast, novel therapeutic strategies should target the epigenetic pathways of synovial activation in RA.

Introduction

Less than a decade ago, the number of genes encoded within the nucleus of a single human cell was estimated to at least 100 000 genes. Much hope for our understanding of pathogenesis and treatment of diseases was put on the successful accomplishment of the human genome project. It was, therefore, most surprising that the number of genes finally detected was quite low. Most of the 25 000 genes identified encode biological functions that remain undiscovered so far, and the functional characterization of these genes in normal physiology as well as in the pathogenesis of diseases remains the main issue for biomedical research in the coming years [1]. However, this approach through "functional genomics" might be biased by post-replicational, post-transcriptional as well as post-translational modifications – and proteome diversity due to alternative splicing of mRNA transcripts and other bio-chemical alterations somehow limits the utility of genomic information [2, 3]. The question how the genome integrates intrinsic and environmental factors thus might be answered by the emerging concept of the "epigenotype". The term "epigenetics" comprises stable alterations of the genetic information that are heritable but do not involve mutations of the DNA sequence itself. Epigenetic regulations are mediated by several biochemical phenomena, most importantly, however, by "histone modi-fications" and "DNA methylation". Epimutations and epigenetics are required for development and differentiation of cells within a multicellular organism. Moreover, they allow a cell to respond to environmental stimuli throughout adult life by means of stable expression or repression of genes in specific cell types. Currently, no epige-netic information is systematically analyzed and epigenetic modifications have not been assessed within the human genome project. Since epigenetics might play the linking role between environmental factors and genetics in determining a certain phenotype, the investigation of epigenetic alterations along the lines of chromo-some-wide and promoter-specific arrays will represent an fascinating area of future research. Already, pilot studies for the human epigenome project have been under-taken [4]. In this regard, the epigenome could provide a readout of an individual's environmental history [5]. The conventional method of studying human diseases by molecular genetic approaches and additional environmental factors could soon be extended to a novel field of "epigenetic epidemiology".

Within this chapter, we focus on the emerging concept of epigenetics and its implications for potential treatment strategies in rheumatoid arthritis (RA).

Definitions

The term epigenetics is used to refer to molecular processes that regulate gene expression patterns, without however changing the DNA nucleotide sequence. The epigenome of a cell is defined by two major groups of biochemical alterations: post-transcriptional methylation of DNA and modifications of histones that package DNA and, thus, modulate the accessibility for transcription factors to information present on nucleic acid. These modifications are mitotically heritable and can be transmitted during cell division from one generation of cells to the next. Both physi-ological and pathological responses to environmental stimuli are probably mediated

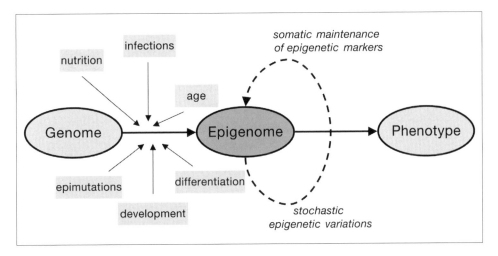

Figure 1

Interaction between genes (genome), epigenetics (epigenome) and phenotype (adapted and modified after [49]). Post-synthetic modifications of DNA are inherited (as epimutations) or establish during development and cell differentiation. Physiological and pathological responses to environmental stimuli (such as nutrition, age, and infections) are also governed by epigenetic mechanisms. Even in the absence of such environmental factors, the epigenomic profile is reversible and highly variable, probably due to stochastic events in the somatic inheritance and in maintenance of epigenetic profiles.

by epigenetic mechanisms. Even in the absence of such environmental factors, the epigenomic profile is reversible and highly variable, probably due to stochastic events in the somatic inheritance of epigenetic profiles [6]. The remarkable degree of variation distinguishes "epimutations" and "epigenetic changes" from "true" genetic mutations. From a therapeutic viewpoint, the reversibility of epigenetic changes finally offers the opportunity of using pharmacological strategies to revolve a certain phenotype [7]. The dynamic interaction between all factors involved is illustrated in Figure 1.

DNA methylation

DNA methylation [8, 9] in eukaryotes is a post-replication modification restricted to the pyrimidine base cytosine within the dinucelotide sequence CpG, which forms clusters of genetic regions (CpG islands) that surround the promoter region of protein-coding genes. Phenotypic analyses of several DNA methyltransferases (DNMTs) have revealed interesting mechanistic insights into epigenetic methylation:

DNMT3a and 3b for example are *de novo* methyltransferases and introduce cytosine methylation at CpG sites that were previously unmethylated. DNMT1 acts during replication and cell division by copying existing methyltransferase pattern onto newly synthesized DNA strands. For the process of DNA methylation, DNMTs use S-adenosylmethionine, whose generation is mainly modulated by the availability of different methyl donors. Briefly, methyltetrahydrofolate governs the conversion of homocysteine to methionine, which is further metabolized to S-adenosylmethionine. Deficiencies in the enzymes involved in these processes result in hypomethylation of DNA. Several nutrients play key roles within this metabolism. The major dietary sources of methyl groups include folate, choline and vitamin B12 [10–13].

The insertion of a methyl group at position five of the cytosine ring leads to structural changes of chromatin and is associated with gene repression. This silencing function on the level of gene expression can be achieved by different mechanisms. Structural modifications of the DNA might block the proper docking of DNA-binding factors to their fitting recognition sites, thus inhibiting gene transcription, whereas methyl-CpG-binding proteins (MBPs, such as MeCP2, MBD1–4) function redundantly as transcriptional co-repressors. Moreover, MBPs have been shown to interact with enzymes regulating histone modifications. This interaction could provide a link between different epigenetic mechanisms.

Histone modifications

Apart from DNA methylation, local chromatin architecture and, thus, transcriptional regulation of gene expression is strongly influenced by covalent biochemical modifications subsumed under the term "histone code". These epigenetic changes include post-transcriptional modifications of histones, including methylation, phosphorylation, ubiquitination, sumoylation, biotinlyation and, most importantly, deacetylation and acetylation [14–16]. "Nucleosomes" are the fundamental building blocks of the heterochromatin consisting of an octamer of four core histones and DNA. This octameric structure is made out of an H3-H4 tetramer and two H2A-H2B dimers. Histone H1 has a linker function between DNA and protein and governs the path of the DNA as it exits from the nucleosome. With respect to histone modification, the DNA of physiologically resting cells is wrapped tightly around the core histones, thus preventing the binding of basal transcription factors (e.g., the TATA box binding protein) and RNA polymerase II [17]. Gene transcription is initiated when histones are modified to create an open, accessible form of chromatin. Probably the most investigated modification of the histone code is the (de)acetylation of core histones. Histone acetylation is performed by histone acetyltransferases (short: histone acetylases, HATs) that neutralize positive charges at the ε amino groups of lysine residues at the N termini. Hyperacetylation of histones is generally associated with enhanced rates of gene transcription. Conversely, the space between histones and surrounding DNA is reduced by de- and hypoacetylation, and transcription factors are sterically hindered from binding, leading to gene silencing. Taken together, the gene transcription rate is regulated by the balance between histone acetylation and histone deacetylation. The targeted deacetylation of histones is performed by several multisubunit enzyme complexes, i.e., histone deacetylases (HDACs) [16, 18]. Figure 2 shows the dynamic interplay of epigenetic mechanisms between states of silent and transcriptionally active chromatin.

Eukaryotic members of the HDAC family can be divided into three major groups, of which class I and class II HDACs so far comprise the best characterized classes with respect to function. The class I comprises HDAC1–3 and 8, and localize almost exclusively within the nucleus to exert their function. HDACs 4–7, 9, and 10 are subsumed under the term class II HDACs, which are mainly found in the

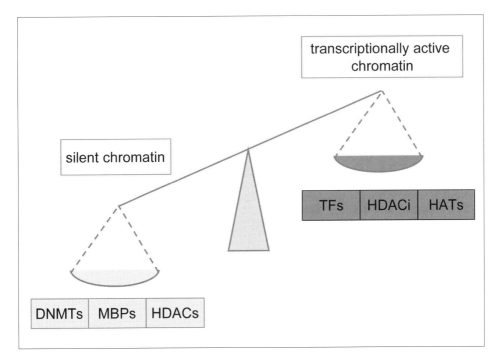

Figure 2
The dynamic balance between silent and transcriptionally active chromatin (modified after [23]). Histone deacetylases (HDACs), DNA methyltransferases (DNMTs) and methyl-CpG-binding proteins (MBPs) provide gene repression, whereas transcription factors (TFs), histone acetyltransferases (HATs) and HDAC inhibitors (HDACi) lead to enhanced gene transcription rates. Disturbances and changes in one or more of these components shift the balance to any side of gene expression.

cytoplasm and are shuttled into the nucleus when needed. Whereas class I HDACs are expressed in most cell types, the expression pattern of class II HDACs appears to be tissue-specific, indicating a possible role in cellular differentiation and development. HDACs remove the acetyl group from the nucleosomal core histones using a sophisticated charge-relay system using Zn^{2+} ions as prosthetic group. HDAC inhibitors such as Trichostatin A (TSA) fit into the active catalytic pocket of the enzyme, exchange the zinc ion and, thus, make the system dysfunctional [14, 19].

HATs, on the other hand, comprise a family of proteins that catalyze the acetylation of lysine residues of one of the core histone proteins [20–22]. Traditionally, HATs are categorized in two groups based on their subcellular localization: type A, located in the nucleus and type B, located in the cytoplasm. Nuclear type A HATs acetylate nucleosomal histones within chromatin in the nucleus, and thus type A HATs are related to transcriptional regulation processes. On the other hand, type B HATs acetylate newly synthesized free histones in the cytoplasm. Since recent data indicate that HAT activity can also be induced in multiple protein complexes that are related to transcriptional processes, this historical categorization is no longer used. Amino acid sequence analyses of all HAT proteins revealed the important feature that HAT proteins fall into distinct families that share relatively poor sequence similarity. All three superfamilies (GNAT, MYST and p300/CBP-HATs), however, have a highly conserved acetyl-CoA binding site in common.

The best-understood family of HATs is the GNAT [general control non-derepressilble-5 (Gcn5)-related N-acetyltransferase] superfamily. Humans express two Gcn5-like acetyltransferases: Gcn5 and PCAF (p300/CBP-associated factor). Both of these proteins can interact with another HAT protein complex, i.e., p300/CBP. P300/CBP is a ubiquitously expressed, global transcriptional co-activator that regulates cell cycle, differentiation and apoptosis. The HAT activities of p300/CBP enables the transactivation of DNA binding transcription factors (p53, E2F, myb, GATA1, Rb) as well as the acetylation of all four histone proteins. Mutations in the HAT active site inhibit their transcriptional activating function.

The MYST family of HATs is particularly interesting as these proteins show similarity with other acetyltransferases exclusively within the acetyl-coenzyme A binding motif. The members of the MYST family in this HAT group are: MOZ (monocytic leukemia zinc finger protein), Ybf2/sas3, Sas2, and Tip60. MOZ is involved in the chromosome translocations associated with acute myeloid leukemia. MOZ acts as a transcriptional coactivator for AML1, which is essential for establishment of definitive hematopoiesis. An overview of the different HAT families is provided in Table 1.

Finally, DNA methylation and histone modifications have been considered as two distinct mechanisms, which influence the level of gene expression in an independent manner. It was shown, however, that HDACs are correlated to DNA methylation, either through direct interaction of HDACS with DNMTs or by the function of MBPs. Another line of evidence suggesting that MBPs interact with HDAC1 and

Table 1. Histone acetyltransferase (HAT) superfamily (modified after [50]).

HAT superfamily	HAT	Transcription-related functions	Histones acetylated	Interaction with other HATs
GNAT	Gcn5	Co-activator	H3/H4	p300; CBP
	Hat1		H4	
	PCAF	Co-activator	H3/H4	p300; CBP
	Elp3	Transcript elongation		
	Hpa2		H3/H4	
MYST	Esa1	Cell cycle progression	H4/H3/H2A	
	MOF		H4/H3/H2A	
	Sas2	Silencing		
	Sas3	Silencing	H3/H4/H2A	
	MORF		H4/H3/H2A	
	Tip60	HIV Tat interaction	H4/H3/H2A	
	Hbo1	ORC interaction		
p300/CBP	P300	Global co-activator	H2A/H2B/H3/H4	PCAF; Gcn5
	CBP		H2A/H2B/H3/H4	PCAF; Gcn5

HDAC2 to recruit the Sin3 corepressor protein further supports a link between DNA methylation and histone modifications [23]. Thus, DNMTs appear to be forms of dual function proteins, which are recruited by transcriptional repressors. On the other hand, they have a non-enzymatic function interacting with histone methyltransferases and HDACs, hence leading to chromatin remodeling [24–26].

Epigenetic modifications in RA

RA is a systemic disease mainly affecting the synovial tissues of joints. The process of ongoing inflammation and erosion of articular cartilage and subchondral bone causes severe pain, functional impairment and ultimately disability [27]. As in other autoimmune-related disorders, neither the etiology nor the pathogenesis of RA has yet been completely unraveled. It is generally accepted, however, that autoimmune disorders develop through a combination of the individual genetic susceptibility, environmental factors, and dysregulated immune responses [28, 29].

Genetic predisposition has been described in RA, in particular as "shared epitope", a distinct sequence of amino acids within the antigen-presenting peptide groove of the major histocompatibility complex. Imbalanced immunity is reflected by the production of autoantibodies such as rheumatoid factors and cyclic citrullinated peptides as well as by the accumulation of reactive helper T cells within the rheumatoid synovium. In addition, environmental factors have been postulated as disease-modulating agents, including smoking, nutrition and infectious agents. So far, these factors have been studied almost exclusively as separate disease-driving agents. As mentioned before, however, the way genes are transcribed can be affected by environment, nutrition, and ageing, which influence the epigenetic patterns.

In this regard, Kim and coworkers [30] investigated whether methotrexate induces genomic DNA hypomethylation in patients with inflammatory arthritis. Methotrexate is an immunosuppressive agent, which is used in the treatment of RA in addition to glucocorticosteroids and biologicals. Methotrexate exerts its function by blocking the enzyme dihydrofolate reductase thus interrupting the methyl transfer function of folate. Within the study in question, the amount of methylated genomic DNA was lowest in subjects with inflammatory arthritis who were not taking methotrexate, highest in subjects with inflammatory arthritis who were taking methotrexate, and intermediate in control subjects. Surprisingly, these data indicate that inflammatory arthritis might be associated with genomic DNA hypomethylation, which can be reversed with methotrexate [30]. Since methotrexate would be expected to inhibit transmethylation, the results from this study are somewhat paradoxical.

In an animal model of adjuvant arthritis, treatment with the combination of tryptophan, methionine and methotrexate caused a drastically reduced course of arthritis, whereas treatment with methotrexate alone exerted only a slight inhibitory effect [31]. The methyl groups transferred during methylation of DNA are ultimately derived from methionine. Therefore, high methionine intake was suggested to increase DNA methylation. Because of the distinct biochemistry within the methionine cycle, methylation of DNA is probably impaired under dietary excess of methionine by inhibiting remethylation of homocysteine [32].

Mature B lymphocytes and plasma cells express CD21 (complement receptor II) on the cell surface to exert their function as (auto-)antibody-producing cells. CD21 binds the complement component C3d, which is expressed within immune complexes. Interestingly, immature precursor of B cells, such as pro-, pre-, or plasma B lymphocytes do not express CD21, probably because they contain a methylated CpG island within its promoter region. When peripheral blood mononuclear cells and synovial fluid mononuclear cells from patients with RA were investigated, however, the CD21-CpG island was found to be demethylated [33]. These data suggest a role of these cells in the dysregulation of the immune response in RA patients.

It is of interest to note here that most therapies have been targeting T and B lymphocytes and/or monocytes/macrophages and their respective proinflammatory cytokines, but never the synovial fibroblast (SF), which was shown to be activated even in the absence of stimulating immune cells and cytokines [34]. This observation has prompted us to study signaling pathways in RA SFs. Since DNA methylation is also involved in the regulation of endogenous retroviral sequences, which have been suggested to play a role in the induction of autoimmune diseases, we searched for the presence of such sequences. By screening RA synovial fluid pellets, a homologue to the human retrotransposable L1 element was found by our group [35]. Human L1s (or LINE-1, long interspersed nuclear elements) are poly(A)-retrotransposons lacking long terminal repeats. L1s contain an untranslated region (UTR) and two open reading frames (ORF1 and 2), the latter encoding for a protein with endonuclease and reverse transcriptase activity. Inhibition of the DNA methyltransfer by 5-aza-2'-deoxycytidine induced the expression of L1s by DNA hypomethylation. It was further shown that three of five CpG islands of the genomic L1 5'-UTR were hypomethylated in RA SFs. Moreover, L1 retroelements induced the expression of p38δ, one of the four isoforms of mitogen-activated protein kinases as well as galectin-3 binding proteins [35, 36].

With respect to histone acetylation, Ito and coworkers [37] showed that reduced activity of class I HDAC enzymes leads to enhanced transcription of genes encoding inflammatory proteins, such as tumor necrosis factor (TNF)-α, interleukin (IL)-8 and matrix metalloproteinase (MMP)-9. These studies were performed in lung tissue derived from patients with inflammatory and obstructive pneumopathies such as asthma bronchiale and chronic obstructive pulmonary disease (COPD). Our laboratory has investigated HDAC activity levels in total synovial tissue of patients with RA [38]. When we compared these samples with the respective levels in patients with osteoarthritis and normal synovial tissue, we found a significant decrease of total HDAC activity. This was probably due to reduced expression of HDAC 1 and 2 proteins in synovial tissue. Conversely, several studies have proposed HDAC inhibitors as beneficial agents for the treatment of RA and other inflammatory processes [39, 40].

It remains unclear whether the observed reduction in HDAC activity is "the chicken or the egg" in the pathogenesis of RA, and it cannot be excluded that the reduced HDAC activity levels reflect an epiphenomenon of ongoing inflammation [38]. Evidence for a key role of histone modifications is based on the fact that the pro-inflammatory transcription factor nuclear factor-kappa B (NF-κB) is highly activated in RA synovial cells (for review see [24]). Since Ito et al. [41] have further shown that HDAC 2 suppresses the NF-κB-mediated gene expression, we speculate that class I HDACs appear to act upstream of NF-κB and other related transcription factors in RA. Thus, HDAC activity appears to play a key role in the pathogenesis of autoimmune-related joint diseases.

Jungel and coworkers [42] have finally shown that co-treatment of RA SFs with the HDAC inhibitor TSA and TNF-related apoptosis inducing ligand (TRAIL) induced programmed cell death (apoptosis) in a synergistic manner. When used alone, TRAIL and TSA exhibited no or only a modest effect.

Regulatory T cells (Tregs) have been thought to be of potential benefit for the treatment of autoimmune diseases, and Foxp3 has been proposed as a master regulator governing both development and function of CD4+ T lymphocytes. In this regard, it was recently shown that the expression of Foxp3 has to be stabilized by complete demethylation of CpG islands within this promoter region to develop a permanent suppressor cell lineage. These findings might be of clinical importance with respect to the therapeutical transfer of Tregs in autoimmune diseases [43].

Taken the data on epigenetic modifications on RA SFs together with the fact that all current therapies, including the novel biologicals, do not result in an ACR70 greater than 60%, and that the RA SF has never been targeted by any therapeutic strategy, future efforts should be given to target the activated RA SFs [44].

Perspectives

The current method of investigating the pathogenesis of diseases includes molecular genetic analyses to identify disease-specific gene sequences on one side and epidemiological approaches to detect potential environmental factors on the other. These efforts have only been partly successful in finding explanations for the complex pathogenesis of autoimmune-related diseases such as RA. Epigenetic modifications thus appear to play pathogenetic key roles in genetically predisposed individuals. Distinct from genetic mutations, which are permanent, epigenetic alterations show an intrinsic plasticity and might well be targeted by pharmacological strategies. Various drugs that modulate the epigenetic reactions in RA have already been tested *in vitro* and in animal models. In particular, targeting DNA methylation and histone acetylation are clearly within future therapeutic prospects. Several epigenetic drugs have also been approved by the FDA, in particular for clinical trials addressing the treatment of malignancies (reviewed in [45]). Nucleoside analogues such as azacytidine as well as the first orally bioavailable inhibitor zebularine inhibit DNA methylation by being incorporated into replicating DNA. These agents have been tested successfully in the treatment of hematological cancers and myelodysplastic syndromes. Another group of DNA methylation inhibitors includes non-nucleoside analogues that inhibit DNMT enzyme activity to exert their function. These drugs are already in clinical use and comprise the anesthetic agent procaine, procainamide (an anti-arrhythmic compound) and the anti-hypertensive drug hydralazine. It has been reported that these agents cause global DNA hypomethylation in cancer cells as well as T lymphocytes in different experimental systems. On the other hand, various HDAC inhibitors are currently in Phase I/II clinical trials. These inhibitors include TSA, suberoylanilide hydroxamic acid (SAHA), valproic acid, phenylbutyrate and others. SAHA, for example, was accepted by the FDA only recently for the treatment of advanced cutaneous T cell lymphoma. Other studies have revealed strong anticancer activities of HDAC inhibitors and feasible pharmacokinetic properties for the treatment of hematological tumors. However, it is important to stress that all inhibitors, which are currently used, block the different HDAC enzymes without any preference for specific isoforms. To provide a more specific approach for future therapies, novel compounds have to be developed.

In addition, the combination of agents that block both DNA methylation and histone modifications might be of therapeutical interest. TSA, which has long been considered as a specific histone deacetylase inhibitor leading to histone hyperacetylation and activation of unmethylated gene sequences, has been shown to induce systemic and replication-independent demethylation of DNA [46]. The results from this study should be taken into account when novel, TSA-related epigenetic drugs are designed.

Finally, dietary recommendations for putative epigenetic drugs as found in green tea, garlic, broccoli and other phytochemical compounds [47, 48] might be an additional option apart from pharmaceutical interventions.

Until then, however, novel biologicals that reverse the epigenetic pattern have to be designed. Moreover, hurdles facing *in vivo* efficacy and toxic side effects have to be overcome. The epigenotype has to be investigated, along the line of chromosome-wide and promoter-specific arrays, especially with respect to the activated RA SFs. This will open the scope for future therapies and might push epigenetic inhibitors as potent agents to treat or prevent disease on an individual basis.

References

1   Huber LC, Distler O (2006) Rheumatology related genes, Identification. In: D Ganten, K Ruckpaul (eds): *Encyclopedic reference of genomics and proteomics in molecular medicine*. Springer, Berlin, 1670–1677

2   Southan C (2004) Has the yo-yo stopped? An assessment of human protein-coding gene number. *Proteomics* 4: 1712–26

3   Little PF (2005) Structure and function of the human genome. *Genome Res* 15: 1759–66

4   Rakyan VK, Hildmann T, Novik KL, Lewin J, Tost J, Cox AV, Andrews TD, Howe KL, Otto T, Olek A et al (2004) DNA methylation profiling of the human major histocompatibility complex: a pilot study for the human epigenome project. *PLoS Biol* 2: e405

5   Whitelaw NC, Whitelaw E (2006) How lifetimes shape epigenotype within and across generations. *Hum Mol Genet* 15 (Spec No 2): R131–7

6   Wong AH, Gottesman II, Petronis A (2005) Phenotypic differences in genetically identical organisms: The epigenetic perspective. *Hum Mol Genet* 14 (Spec No 1): R11–8

7   Ballestar E, Esteller M, Richardson BC (2006) The epigenetic face of systemic lupus erythematosus. *J Immunol* 176: 7143–7

8   Klose RJ, Bird AP (2006) Genomic DNA methylation: The mark and its mediators. *Trends Biochem Sci* 31: 89–97

9   Karouzakis E, Neidhart M, Gay RE, Gay S (2006) Molecular and cellular basis of rheumatoid joint destruction. *Immunol Lett* 106(1): 8–13

10 Niculescu MD, Zeisel SH (2002) Diet, methyl donors and DNA methylation: Interactions between dietary folate, methionine and choline. *J Nutr* 132: 2333S–2335S

11 Poirier LA (2002) The effects of diet, genetics and chemicals on toxicity and aberrant DNA methylation: An introduction. *J Nutr* 132: 2336S-2339S

12 Zhu WG, Lakshmanan RR, Beal MD, Otterson GA (2001) DNA methyltransferase inhibition enhances apoptosis induced by histone deacetylase inhibitors. *Cancer Res* 61: 1327–33

13 Januchowski R, Dabrowski M, Ofori H, Jagodzinski PP (2007) Trichostatin A down-regulate DNA methyltransferase 1 in Jurkat T cells. *Cancer Lett* 246: 313–7

14 de Ruijter AJ, van Gennip AH, Caron HN, Kemp S, van Kuilenburg AB (2003) Histone deacetylases (HDACs): Characterization of the classical HDAC family. *Biochem J* 370: 737–49

15 Adcock IM, Ford P, Barnes PJ, Ito K (2006) Epigenetics and airways disease. *Respir Res* 7: 1–21

16 Verdin E, Dequiedt F, Kasler HG (2003) Class II histone deacetylases: Versatile regulators. *Trends Genet* 19: 286–93

17 Barnes PJ, Adcock IM, Ito K (2005) Histone acetylation and deacetylation: Importance in inflammatory lung diseases. *Eur Respir J* 25: 552–63

18 Glozak MA, Sengupta N, Zhang X, Seto E (2005) Acetylation and deacetylation of non-histone proteins. *Gene* 363: 15–23

19 Vigushin DM, Coombes RC (2002) Histone deacetylase inhibitors in cancer treatment. *Anticancer Drugs* 13: 1–13

20 Sterner D, Berger S (2000) Acetylation of histones and transcription-related factors. *Microbiol Mol Biol Rev* 64(2): 435–59

21 Marmorstein R (2001) Structures of histone acetyltransferases. *J Mol Biol* 311(3): 433–44

22 Grabiec AM, Tak PP, Reedquist KA (2008) Targeting histone deacetylase activity in rheumatoid arthritis and asthma as prototypes of inflammatory disease: should we keep our HATs on? *Arthritis Res Ther* 10(5): 226

23 Zhang Y, Fatima N, Dufau ML (2005) Coordinated changes in DNA methylation and histone modifications regulate silencing/derepression of luteinizing hormone receptor gene transcription. *Mol Cell Biol* 25: 7929–39

24 Kishikawa S, Ugai H, Murata T, Yokoyama KK (2002) Roles of histone acetylation in the Dnmt1 gene expression. *Nucleic Acids Res* (Suppl) 209–10

25 Januchowski R, Dabrowski M, Ofori H, Jagodzinski PP (2007) Trichostatin A down-regulate DNA methyltransferase 1 in Jurkat T cells. *Cancer Lett* 246: 313–7

26 Jaenisch R, Bird A (2003) Epigenetic regulation of gene expression: How the genome integrates intrinsic and environmental signals. *Nat Genet* 33 (Suppl): 245–54

27 Huber LC, Gay RE, Gay S (2006) Synovial activation. In: A Falus (ed): *Immunogenomics and Human Disease*. Wiley, New York, 299–325

28 Ermann J, Fathman CG (2001) Autoimmune diseases: Genes, bugs and failed regulation. *Nat Immunol* 2: 759–61

29 Smith JB, Haynes MK (2002) Rheumatoid arthritis – A molecular understanding. *Ann Intern Med* 136: 908–22

30 Kim YI, Logan JW, Mason JB, Roubenoff R (1996) DNA hypomethylation in inflammatory arthritis: Reversal with methotrexate. *J Lab Clin Med* 128: 165–72

31 Kroger H, Dietrich A, Gratz R, Wild A, Ehrlich W (1999) The effect of tryptophan plus methionine, 5–azacytidine, and methotrexate on adjuvant arthritis of rat. *Gen Pharmacol* 33: 195–201

32 Waterland RA (2006) Assessing the effects of high methionine intake on DNA methylation. *J Nutr* 136: 1706S–1710S

33 Schwab J, Illges H (2001) Silencing of CD21 expression in synovial lymphocytes is independent of methylation of the CD21 promoter CpG island. *Rheumatol Int* 20: 133–7

34 Muller-Ladner U, Kriegsmann J, Franklin BN, Matsumoto S, Geiler T, Gay RE, Gay S (1996) Synovial fibroblasts of patients with rheumatoid arthritis attach to and invade normal human cartilage when engrafted into SCID mice. *Am J Pathol* 149: 1607–15

35 Neidhart M, Rethage J, Kuchen S, Kunzler P, Crowl RM, Billingham ME, Gay RE, Gay S (2000) Retrotransposable L1 elements expressed in rheumatoid arthritis synovial tissue: Association with genomic DNA hypomethylation and influence on gene expression. *Arthritis Rheum* 43: 2634–47

36 Kuchen S, Seemayer CA, Rethage J, von Knoch R, Kuenzler P, Beat AM, Gay RE, Gay S, Neidhart M (2004) The L1 retroelement-related p40 protein induces p38delta MAP kinase. *Autoimmunity* 37: 57–65

37 Ito K, Ito M, Elliott WM, Cosio B, Caramori G, Kon OM, Barczyk A, Hayashi S, Adcock IM, Hogg JC, Barnes PJ (2005) Decreased histone deacetylase activity in chronic obstructive pulmonary disease. *N Engl J Med* 352: 1967–76

38 Huber LC, Brock M, Hemmatazad H, Giger OT, Moritz F, Trenkmann M, Distler JH, Gay RE, Kolling C, Moch H et al (2007) Histone deacetylase/acetylase activity in total synovial tissue derived from rheumatoid arthritis and osteoarthritis patients. *Arthritis Rheum* 56: 1087–93

39 Chung YL, Lee MY, Wang AJ, Yao LF (2003) A therapeutic strategy uses histone deacetylase inhibitors to modulate the expression of genes involved in the pathogenesis of rheumatoid arthritis. *Mol Ther* 8: 707–17

40 Blanchard F, Chipoy C (2005) Histone deacetylase inhibitors: New drugs for the treatment of inflammatory diseases? *Drug Discov Today* 10: 197–204

41 Ito K, Yamamura S, Essilfie-Quaye S, Cosio B, Ito M, Barnes PJ, Adcock IM (2006) Histone deacetylase 2-mediated deacetylation of the glucocorticoid receptor enables NF-kappaB suppression. *J Exp Med* 203: 7–13

42 Jungel A, Baresova V, Ospelt C, Simmen BR, Michel BA, Gay RE, Gay S, Seemayer CA, Neidhart M (2006) Trichostatin A sensitises rheumatoid arthritis synovial fibroblasts for TRAIL-induced apoptosis. *Ann Rheum Dis* 65: 910–2

43 Floess S, Freyer J, Siewert C, Baron U, Olek S, Polansky J, Schlawe K, Chang HD, Bopp T, Schmitt E, Klein-Hessling S, Serfling E, Hamann A, Huehn J (2007) Epigenetic control of the foxp3 locus in regulatory T cells. *PLoS Biol* 5: e38

44 Huber LC, Distler O, Tarner I, Gay RE, Gay S, Pap T (2006) Synovial fibroblasts: Key players in rheumatoid arthritis. *Rheumatology (Oxford)* 45: 669–75

45   Lu Q, Qiu X, Hu N, Wen H, Su Y, Richardson BC (2006) Epigenetics, disease, and therapeutic interventions. *Ageing Res Rev* 5: 449–67

46   Ou JN, Torrisani J, Unterberger A, Provencal N, Shikimi K, Karimi M, Ekstrom TJ, Szyf M (2007) Histone deacetylase inhibitor Trichostatin A induces global and gene-specific DNA demethylation in human cancer cell lines. *Biochem Pharmacol* 73: 1297–307

47   Myzak MC, Ho E, Dashwood RH (2006) Dietary agents as histone deacetylase inhibitors. *Mol Carcinog* 45: 443–6

48   Dashwood RH, Myzak MC, Ho E (2006) Dietary HDAC inhibitors: Time to rethink weak ligands in cancer chemoprevention? *Carcinogenesis* 27: 344–9

49   Feil R (2006) Environmental and nutritional effects on the epigenetic regulation of genes. *Mutat Res* 600: 46–57

50   Roth SY, Denu JM, Allis CD (2001) Histone acetyltransferases. *Annu Rev Biochem* 70: 81–120

Perspectives in targeted therapy

Edward C. Keystone

University of Toronto, Division of Rheumatology, Rebecca MacDonald Centre for Arthritis and Autoimmune Diseases, 60 Murray Street, Toronto, ON M5T 3L9, Canada

Abstract

Targeted therapeutic agents have changed the landscape of therapy in rheumatoid arthritis (RA). They have also provided valuable insights into the utility of animal models for development of targeted therapies, clinical trial design, pharmacodynamics, immunobiology and key pathogenic elements of disease. Studies of chimeric anti-CD4 monoclonal antibodies in RA demonstrated the need for pre-clinical studies to more closely approximate the human therapeutic paradigm as well as the importance of synovium as an appropriate pharmacodynamic window to predict efficacy and adverse side effects of the agents. Targeted therapies have been instructive in discerning the importance of TNF, IL-1, IL-6, IL-15 and RANKL in the pathological process themselves, such as the uncoupling of inflammation and structural damage. Current trends in the use of targeted therapeutics include aggressive earlier use, combination with methotrexate, use in moderate rather than severe disease, tight control as well as induration and maintenance regimes. Despite therapeutic advances with target therapies a number of unmet needs exist, including a low remission rate, cost and inadequate access as well as the lack of biomarkers to predict response and safety concerns. Despite this, target therapies have revolutionized the treatment of RA. In addition to having a substantial effect on clinical outcomes, a number of valuable lessons have been learned.

Introduction

An improved understanding of the pathogenesis of rheumatoid arthritis (RA) coupled with recent advances in biotechnology has led to selective targeting of the pathogenic elements of disease utilizing biological agents. As a consequence of the development of targeted agents there has been an explosion in the number of disease modifying agents (DMARDS) approved for the treatment of RA. Prior to 15 years ago, a new DMARD was introduced about every 15 years. Over the last 15 years 8 DMARDS have reached the marketplace – 6 of which are biological targeted therapies. Studies of these biological agents have provided extremely valuable insights into the utility of animal models for development of targeted therapies, clinical trial study design, pharmacodynamics human immunobiology, and key pathogenic elements of disease.

Insights

Proof of efficacy of therapeutic agents in animal models of RA is used to predict efficacy in human disease. Despite this, there are numerous instances in RA where pre-clinical data were not reflective of the human situation. Agents demonstrating benefit in rodent models of RA, but not in RA, include anti-ICAM-1 and anti-IL-8 monoclonal antibodies (mAbs), as well as immunomodulators IL-4, IL-10 and IL-11 (reviewed in [1]). One notable example of the discord between pre-clinical studies and results in humans was the anti-CD4 T cell-depleting mAb. Despite efficacy in several pre-clinical models of RA, anti-CD4 mAbs were shown not to be beneficial in RA [2–5]. However, a number of valuable insights were gained into the pharmacodynamics of immune cell depletion with anti-CD4 mAbs that were not predicted by the animal models. Thus, although profound depletion of circulating $CD4^+$ T cells was observed, synovial T cells were still detected, suggesting that the pharmacodynamic window correlating with therapeutic effect was the synovium and not the circulation [6]. This concept was supported by the positive correlation between anti-CD4 mAb coating of synovial fluid cells, but not circulating $CD4^+$ T cells, with therapeutic benefit [7].

T cell-depletion studies appeared to demonstrate substantial differences in T cell biology in mouse and humans. Thus, despite short-term depletion of $CD4^+$ T cells in rodent models, profound long-term depletion of circulating $CD4^+$ T cells was observed in humans [8]. Of significance, the prolonged T cell depletion observed with anti-CD4 mAbs was actually predicted by pre-clinical data. Thus, whereas short-term $CD4^+$ T cell depletion was demonstrated with an anti-CD4 mAb in young mice, prolonged depletion of $CD4^+$ T cells was observed in older mice with an age comparable to that of mAb-treated RA patients. In addition, pre-clinical models showed more T cell cytotoxicity with a chimeric mAb than with a heterologous counterpart . Thus, chimeric anti-CD4 mAb studies in animal models of RA appeared to reflect the human situation and may have predicted the prolonged T cell depletion in RA. The results provide an important insight into the need for pre-clinical studies to more closely approximate the human therapeutic paradigm.

The importance of an appropriate pharmacodynamic window was emphasized by the failure to observe an increase in infection or malignancy in RA patients despite a profound long-term depletion of circulating $CD4^+$ T cells well below the levels observed in HIV. Nowhere was the failure of pre-clinical studies predictive of an effect on humans more evident than in the multi-organ failure that resulted from infusion of an anti-CD28 mAb in a recent Phase I study [9].

As a consequence of the therapeutic failure of depleting anti-CD4 mAb in RA, non-depleting anti-CD4 mAbs were evaluated. Initial studies were carried out with a primatized IgG4 mAb in which the first generation mAb demonstrated good clinical benefit [10]. A second trial with the same mAb generated by a different manufacturing process yielded significantly reduced efficiency and caused CD4 T cell depletion

[11]. Although both antibodies were generated in Chinese hamster ovary cell lines, differences in the level of aggregation and a nonglycosylated heavy chain were thought to account for the results. The data demonstrate how subtle differences in biological agents can result in substantial differences in pharmacodynamics and clinical effectiveness.

The lessons learned in CD4[+] T cell depletion in RA have been particularly helpful in allaying concerns about B cell depletion. Despite profound depletion of peripheral B cells with rituximab, an anti-CD20 mAb, the incidence of infection appears to be no greater than that observed with tumor necrosis factor inhibitors (TNFi) [12]. Additional insight into the relationship between circulating immune cell depletion and clinical efficacy was shown by the lack of correlation between loss of efficacy after a course of rituximab and repletion of circulating B cells [13]. However, when a high-throughput FACS analysis was utilized, a significant correlation was observed between B repletion and RA flare. The presence of B cells within the synovium despite marked depletion of circulating B cells again emphasizes the concept of the synovium as an important pharmacodynamic window [14]. An analysis of synovium after rituximab treatment demonstrated a significant positive correlation between reduction of clinical disease and initial macrophages and plasma cells [15]. The results suggest that treatment with rituximab causes an indirect decrease in inflammatory cells other than B cells, thus providing insight into the role of B cells orchestrating synovial inflammation. The correlation between the clinical response and synovial plasma cells suggests that rituximab exerts its effect, at least in part, through an effect on autoreactive plasma cells associated with autoantibody production. Other important insights have been gained from evaluation of synovial biomarkers in clinical trials of targeted therapies [16]. One of the most instructive lessons was derived from synovial biopsies following TNF blockade. Although the rapid reduction in cellularity at the site of synovial inflammation with TNF blockade was hypothesized to result from apoptosis of TNF-reactive cells, the clinical efficacy of certolizumab (a PEGylated Fab fragment of a humanized anti-TNF antibody that is unable to induce apoptosis) in RA suggests that induction of apoptosis might not be a requirement for efficacy of TNF inhibitors in RA [17]. More recent synovial biopsy data suggest that rapid reduction in cellularity and inflammation in the rheumatoid synovium after TNF blockade is a result of dampening of TNF-driven cytokine and chemokine cascades associated with a reduction in cellular recruitment and retention of inflammatory cells [18].

Synovial biopsy studies have also provided a rationale for new therapeutic targets in RA. Thus, the accumulation of CCR1 in the synovium of RA patients provided a rationale for CCR1 blockade. A small proof of concept study with an oral CCR1 antagonist demonstrated a marked decrease in the number of macrophages and CCR1-positive cells in actively treated patients associated with a trend towards clinical improvement even after short-term treatment [19]. To date, it is unclear whether blocking a single chemokine receptor is significant to ameliorate RA.

Taken together, the data suggest that analysis of synovial biomarkers can be used for screening purposes during early drug development.

Targeted therapies have also provided insight into the effect of immunomodulation on different immune compartments. Several combinations of biological agents have shown to be ineffective in modulating synovitis at the local tissue level [20, 21]. Yet, substantial effects have been observed on the systemic immune system as indicated by a significant increase in serious infectious and in some cases malignancy [20, 21]. These observations point out the sensitivity of the systemic immune system to immunodulation compared with the local immunoreactive site such as the inflamed joint. The results suggest significant caution is necessary when considering a strategy using a combination of targeted therapies for human disease.

As with efficacy, some adverse events may have been predicted by animal models. Pre-clinical data might have predicted susceptibility of TNFi-treated patients to intracellular infections such as tuberculosis (Tb). Several studies of TNFi in a latent Tb model in mice clearly demonstrated susceptibility for Tb to disseminate [22]. This raises the issue of the diligence required by industry to determine risks prior to treatment of human disease. In addition, more pre-clinical studies involving challenge with infectious agents known to be controlled by the targeted immune element are needed.

Selective targeting of TNF has also provided preliminary evidence for the role of TNF in tumor surveillance. Despite a substantial body of data demonstrating that immune modulation is not generally associated with non-cutaneous solid tumors, studies with TNFis have challenged this notion. The increase in solid malignancies observed in association with the combination of the subtle TNF receptor (etanercept) and cyclophosphamide as well as high-dose infliximab has been sobering [23]. The results of the metanalysis of anti-TNF mAbs by Bongartz, although controversial, have enhanced our index of suspicion for the possibility that TNFi may increase the risk of solid tumors [24].

Some adverse events of targeted therapies may not be predicted in the animal models of disease. Recent reports of multifocal leukoencephalopathy with anti-VCAM-1 (natalizumab) in patients with multiple sclerosis [25] and with rituximab in patients with systemic lupus erythematosus [26] highlight this concept. They emphasize the need for long-term safety surveillance, to detect rare adverse events. Although numerous long-term databases of biologicals have been established across the globe, differences in criteria to initiate a biological, the surveillance data collected, validation and monitoring techniques, and particularly the nature of the non-biological controls make comparisons and/or metanalyses difficult. While propensity score have improved the comparability of control populations, more attention must be paid to disease severity (i.e., risk of damage over time) and disease activity over time. More precise data with respect to the reason for failure of a DMARD or biological, i.e., efficacy *versus* safety, and the adequacy of a treatment course must be taken into consideration. Data on progression of structural damage, genomics and biomarkers in conjunction with high quality clinical data are needed.

A recent review of the lessons learned from therapies with biologicala has emphasized the role of FcγR receptors in predicting differences in the outcome of therapies having the same target (reviewed in [27]). The authors note that differences in efficacy and adverse event profiles of mAbs can be predicted based on FcγR polymorphisms. Additionally, they point out the association of disease and infectious susceptibility with specific FcγR polymorphisms. Although target antigen specificity is the major influence on the benefit risk profile of therapies, the consequences of mAb binding are influenced Fc design. The authors conclude that disease indication should determine Fc design and that collecting information in clinical trials of FcδR genotype would be useful.

One of the most valuable attributes of targeted therapies is the dissection of key pathogenic elements of disease. Therapeutic studies have clearly delineated some of the key cytokine mediators in the pathogenesis of RA, including TNF [28–30], IL-6 [31], and RANKL [32]. Disorders such as psoriasis, psoriatic arthritis, colitis and sarcoidosis have also benefited substantially by selective targeting. The role of T and B cells in the pathogenesis of RA has been clarified through the clinical benefit derived from CTLA4 Ig [33] and anti-CD20 mAb [34], respectively.

When a particular molecular entity is addressed by a targeted therapeutic agent, significant insights have been provided into the pharmacodynamic/pharmacokinetic properties of the agents as well as disease pathogenesis. Thus, differential clinical benefit has been observed with agents targeting the same molecular entity. While anti-TNF mAbs, infliximab and adalimumab as well as etanercept, a soluble TNFR-Fc fusion protein demonstrated clinical benefit in RA, only the anti-TNFm Abs demonstrated benefit in Crohn's disease, sarcoidosis, and Wegener's granulomatosis [35]. Studies are currently underway to address mechanisms accounting for these clinical differences to gain a better understanding of the pathogenesis of these disorders. The efficacy of a PEGylated humanized Fab' fragment targeting TNF in Crohn's disease provides additional insights into the mechanism of action of TNFi as well as disease pathogenesis [36]. A striking example of the insight provided by targeted therapy on the pathogenetics of disease is the amelioration, in part, of multiple sclerosis by rituximab – a B cell-depleting agent [37]. While the rationale for the use of a B cell-depleting agent in patients with multiple sclerosis (previously considered a T cell driven disease) is unclear, the result suggests that classifying diseases as either T cell or B cell directed may be all too naïve. Together these data emphasize the utility of targeted therapeutic approaches to dissecting the pathogenic elements of disease.

Immunogenicity of targeted biological therapies was expected but has still generated a few surprises. The relatively low immunogenicity of chimeric mAbs (i.e., infliximab) particularly with methotrexate (MTX) has been particularly rewarding [38]. Over time and with an increased frequency of infusion, reactions particularly

211

with monotherapy or after a time gap between infusions have been observed [39]. The reason why one TNFi, infliximab, reduces a sustained efficacy while another, adalimumab, reduces initial efficacy is still not clear. A reduction in human anti-chimeric antibodies has been accomplished though the use of high doses of TNFi and regular periodicity of infusions. It is of interest that despite IgG Fc components of soluble receptors, such as TNFR-Fc (etanercept) or CTLA4 Ig (abatacept), little effect of immunogenicity on efficacy or safety has been reported. More surprising was the generation of neutralizing antibodies against the "fully human" anti-TNF mAb, adalimumab. These antibodies against adalimumab are associated with lower serum adalimumab concentrations and a lack of response to adalimumab treatment [40]. Taken together, the data provide a clear signal that the efficacy of any mAb, regardless of the degree of humanization, is likely to be influenced by immunogenicity.

The development of biologicals has had a significant effect on clinical trial design. Anti-CD4 mAb therapy was one of the first unique approaches to RA treatment. Although eight open label studies demonstrated promising results in 60–75% of patients, randomized placebo-controlled trials of both murine and chimeric anti-CD4 mAbs demonstrated no clinical efficacy (reviewed in [41]). The results likely reflected an expectation bias on the part of the investigator and patients [42]. As a consequence of these observations, Phase I trials were blinded thereafter to avoid this pitfall.

Targeted therapies have been instructive not only in discerning the elements that are pathogenic in RA but have also in defining the mechanisms involved in the pathological processes themselves. One striking example of the latter was the discovery of a significant uncoupling of inflammation (as determined by joint swelling) and structural damage (assessed radiographically) particularly with biologicals. Numerous studies have demonstrated that, despite a similar degree of joint swelling in patients treated with MTX or a TNFi, striking differences in radiographic progression are seen [43–45]. Moreover, TNFis have demonstrated superior inhibition of radiographic progression at every level of response and disease activity state achieved. The minimal radiographic progression that occurs in TNFi-treated patients in a low disease activity state, in contrast to MTX where progression continues, suggests the possibility that a low disease activity state in patients that are comfortable with their symptoms may be an acceptable therapeutic target. In patients receiving MTX alone, remission would still be the only acceptable target. Future studies evaluating a low disease activity state in terms of progressive deformity, disability and cardiovascular outcomes are needed. The data suggest that the pathogenic elements leading to inflammation are not identical to those leading to structural damage in the joints. It is conceivable based on these data that TNF is a critical cytokine in causing structural damage, while cytokines in addition to TNF play a significant role in generating inflammation. Further studies utilizing more sensitive imaging techniques, such as ultrasound and MRI, may provide further insight into the observed dissociation. The capability of selective targeting of the elements leading to structural damage in RA has also shown us that healing of damage is possible.

Recent studies of early use of targeted biological agents suggest the possibility of a window of opportunity in RA to change the course of disease. Thus, virtually all patients achieving low disease activity with the initiation of combination of MTX and infliximab were able to continue on low-dose MTX after discontinuation of infliximab for up to 3 years [46]. A significant caveat to the concept of induction and maintenance is the recent demonstration of an inability to discontinue infliximab in patients with baseline characteristics suggesting a poor prognosis [47]. These findings suggest that only patients who had a good prognosis and were likely to respond well to MTX were able to discontinue infliximab. More importantly, 26% of patients achieving low disease activity were able to discontinue all DMARDs. This contrasts with the inability of patients initiating MTX only to discontinue the drug even when progression from undifferentiated arthritis to RA is prevented [48].

Current and future treatment strategies

TNFis have dramatically changed the therapeutic paradigms in RA. They have resulted in markedly improved clinical outcomes with a substantial reduction in irreversible structural damage. As a consequence of TNFi, the therapeutic goal in RA of complete remission including clinical, laboratory and imaging outcomes is now achievable. The success of TNFi has led to their earlier and increased use, estimated in the USA to constitute as much as 35% of DMARD-treated RA patients. Longer term experience with efficacy and safety of TNFi will undoubtedly increase their use further. Currently, the therapeutic paradigm in the treatment of RA is the initiation of MTX to a dose of 15–25 mg for 3–6 months followed by initiation of a TNFi in MTX inadequate responders. If an adequate response is not achieved, the trend is to switch to another TNFi, before the use of a newly approved biological such as abatacept or rituximab is recommended.

A significant trend in clinical practice where TNFis are readily accessible is their use in patients who have more moderate disease activity unlike those in clinical trials where the majority of patients had severe disease activity. Retrospective analyses of clinical trial patients as well as those in surveillance databases of clinical practice with moderate RA have shown marked improvement in the outcomes of patients moderate RA patients, with a considerably larger proportion of patients achieving a low disease activity or remission state compared with patients who initially had severe disease activity [49]. Of note, the methodology used to determine response to therapy, i.e. ACR responses, significantly underestimates the proportion of patients achieving a low disease activity state. Thus, ACR20 non-responders may have substantial reductions in their ACR core set measures such as tender and swollen joints, ESR etc. This likely accounts for some of the dissociation between clinical and radiological outcomes in RA. Taken together, the results suggest that patients in clinical practice who are treated with TNFis have a much better outcome than those evaluated in clinical trials. The data suggest then that the outcomes in clinical trials do not reflect those in clinical practice. These results have important implications for the field. The data suggest that pharmacoeconomic analyses of TNFi should be performed on patients initiating therapy with moderate RA to more closely approximate the situation in clinical practice. Such analyses will likely show a more marked pharmacoeconomic benefit than has been observed with clinical trial population. This could substantially enhance the access to TNFis by the payers. The data also suggest that the unmet need for new therapies in the context of TNFi availability may not be as great as previously thought. A concept that is likely to drive biological use is that of an imaging remission in addition to clinical remission. Improved imaging technology (MRI and ultrasound) has resulted in better detection of synovitis compared with the clinical exam. Increased use of these techniques for detecting synovitis coupled with a greater expectation on the part of rheumatologists to achieve a remission will increase the tendency to aggressively treat patients even in a low disease activity state.

Given their substantial clinical and radiographic efficacy, current TNFis will remain the first line targeted therapy. They have raised the bar for more agents currently in development. This concept is supported by the second line use of both abatacept and rituximab: after TNFis-abatacept because of a more modest radiographic outcome, and rituximab because of safety issues associated with retreatment and use of another biological in the case of rituximab failures while B cells remain depleted.

With the realization that rapid progression of structural damage occurs in early RA, a number of novel therapeutic strategies have been developed to reduce such progression. The utility of initiating a TNFi and MTX early in disease has demonstrated unequivocally the superiority of combination therapy over monotherapy. Unfortunately, the pharmacoeconomic data to support its use in very early RA are still inadequate. The use of tight control strategies has escalated in recent years. Its beneficial affect has been demonstrated but issues such as time to optimal response and target endpoints require further clarification. Although substantial data support TNFi switching, a number of confounders mar their interpretation. These include the small sample size, short duration of studies, poorly defined outcomes, the single center studies and most particularly, the lack of controlled studies. The issues above coupled with inadequate information concerning the optimization of dose and duration of prior TNFi therapy confounds the interpretation of the studies. Controlled trials are clearly warranted.

The future of targeted therapies in RA over the next 10 years is difficult to predict. However, a number of factors will drive utilization. A key driver will continue to be cost and hence access to expensive biologicals. The increasing co-payment for agents funded by private payers, coupled with rising costs will dominate any discussion of utilization. Over the next several years, approval is expected of novel agents interfering with new targets such as anti-IL-6 receptor mAb (tocilizimab) and anti-RANKL mAb (denosumab). The results of several novel B cell-depleting agents are encouraging. Whether low-cost TNFis can be generated with small molecule agents or a PEGylated anti-TNF-directed Fab fragment, i.e., certolizumab, by cheaper technology remains unclear. The failure to develop small molecule inhibitors of signal transduction (i.e., p38) has been discouraging; however, recent preliminary data from studies of the small molecule inhibitor of JAK3 have generated excitement in the field. Other targets in development that are currently being evaluated include IL-15, IL-17 and vascular endothelial growth factor, and B cell-stimulating factors, i.e., BLYSS and APRIL.

Despite therapeutic advances with targeted therapies, there are a number of unmet needs. Although many patients now achieve a low disease activity state with the new therapeutic regimens, fewer achieve complete remission. Sensitive imaging techniques suggest that these patients still have subclinical synovitis with the potential for further structural damage. Further research is need to define the imaging thresholds for progressive structural damage and loss of function. Cost remains a barrier to accessing current targeted therapies, and is a key concern; what is even more worrisome is the reluctance of the majority of rheumatologists to embrace the use of targeted therapies despite their excellent risk/benefit profile. Even ~10 years after approval by regulatory authorities, 80% of biologics are prescribed by ~20% of rheumatologists even where access is not a barrier to use. Whether this imbalance in prescribing habits reflects over aggressive therapy by early adopters or under-utilization by slow adopters needs to be clarified. A critical unmet need with respect to agents where cost and access is an issue is the need for biomarkers to predict appropriate utilization in the patient based on efficacy and safety concerns. The imprecision of physician-driven outcomes and modest correlation with long-term structural damage and disability supports the need for surrogate markers for both disease activity and severity. While many biomarkers are available, few surrogates exist.

In summary, targeted therapies have made an enormous impact in the field of rheumatology, particularly for RA. They have substantially reduced signs and symptoms of disease, improved function and quality of life, and prevented structural damage and hence disability. They have been hugely instructive in the immunology of disease and normal human biology. With the evolution of small molecule inhibitors, further advances in biotechnology and a better understanding of the intricacies of the pathogenic processes in RA, there should be a time when cost is no longer the driver for these novel therapies and access is no longer the barrier to complete remission.

Acknowledgement

The author acknowledges the excellent administrative assistance of Ms. Correna LeCoure.

References

1 Keystone EC (2003) Abandoned therapies and unpublished trials in rheumatoid arthritis. *Curr Opin Rheumatol* 15: 253–258

2 Moreland LW, Pratt PW, Mayes MD, Postlethwaite A, Weisman MH, Schnitzer T, Lightfoot R, Calabrese L, Zelinger JJ, Woody N (1995) Double-blind placebo-controlled multicentre trial using chimeric monoclonal anti-CD4 antibody, CM-T412 in rheumatoid arthritis patients receiving concomitant methotrexate. *Arthritis Rheum* 38: 1581–8

3 Van der Lubbe PA, Dijkmans BA, Markusse HM, Nassander V, Breedveld FC (1995) A randomized double-blind placebo controlled study of CD4 monoclonal antibody therapy in early rheumatoid arthritis. *Arthritis Rheum* 38: 1097–1106

4 Horneff G, Burmester GR, Emmrich F, Kalden JR (1991) Treatment of rheumatoid arthritis with anti-CD4 monoclonal antibody. *Arthritis Rheum* 34: 129–140

5 Tak PP, van der Lubbe PA, Cauli A, Daha MR, Smeets TJ, Kluin PM, Meinders AE, Yanni G, Panayi GS, Breedveld FC (1995) Reduction of inflammation after anti-CD4 antibody treatment in early rheumatoid arthritis. *Arthritis Rheum* 38: 1456–1465

6 Ruderman EM, Weinblatt ME, Thurmond LM, Pinkus GS, Gravallese EM (1995) Synovial tissue response to treatment with CAMPA TH-IH. *Arthritis Rheum* 38: 254

7 Choy EH, Pitzalis C, Cauli A, Bijl JA, Schantz A, Woody J, Kingsley GH, Panayi GS (1996) Percentage of anti-CD4 monoclonal antibody-coated lymphocytes in the rheumatoid joint is associated with clinical improvement. *Arthritis Rheum* 39: 52–56

8 Moreland LW, Pratt PW, Bucy RP, Jackson BS, Feldman JW, Koopman WJ (1994) Treatment of refractory rheumatoid arthritis with a chimeric anti-CD4 monoclonal antibody. Long term follow up of CD4+ T-cell counts. *Arthritis Rheum* 37: 854–858

9 Suntharalingam, G. Perry MR, Ward S, Brett SJ, Castello-Cortes A, Brunner MD, Panoskaltsis N (2006) Cytokine storm in a phase 1 trial of the anti-CD28 monoclonal antibody TGN1412. *N Engl J Med* 355, 1018–1028

10 Levy R, Weisman M, Wiesnehutter C et al (1996) Results of a placebo-controlled multicentre trial using a primatized non-depleting, anti-CD4 monoclonal antibody in the treatment of rheumatoid arthritis. *Arthritis Rheum* (Suppl): 122

11 Mason U, Aldric J, Breedveld F, Davis CB, Elliott M, Jackson M, Jorgensen C, Keystone E, Levy R, Tesser J (2002) CD4 coating, but not CD4 depletion, is a predictor of efficacy with primatized monoclonal anti-CD4 treatment of active rheumatoid arthritis. *J Rheumatol* 29, 220–229

12 Keystone E, Fleischmann R, Emery P, Furst DE, van Vollenhoven R, Bathon J, Dougados M, Baldassare A, Ferraccioli G, Chubick A (2007) Safety and efficacy of additional

courses of rituximab in patients with active rheumatoid arthritis: An open-label extension analysis. *Arthritis Rheum* 56: 3896–3908

13 Emery P, Breedveld F, Martin-Mola E, Pavelka K, Szczepanski L, Kim D, Magrini F, Behrendt C, Kelman A (2006) Relationship between peripheral B Cell return and loss of EULAR response in RA patients treated with rituximab. *Arthritis Rheum* 54: S66

14 Vos K, Thurlings RM, Wijbrandts CA, van Schaardenburg D, Gerlag DM, Tak PP (2007) Early effects of rituximab on the synovial cell infiltrate in patients with rheumatoid arthritis. *Arthritis Rheum* 56: 772–8

15 Thurlings RM, Vos K, Wijbrandts CA, Zwinderman AH, Gerlag DM, Tak PP (2008) Synovial tissue response to rituximab: Mechanism of action and identification of biomarkers of response. *Ann Rheum Dis* 67: 917–25

16 Gerlag DM, Tak PP (2008) Novel approaches for the treatment of rheumatoid arthritis: Lessons from the evaluation of synovial biomarkers in clinical trials. *Best Pract Res Clin Rheumatol* 22: 331–323

17 Keystone EC, van der Heijde D, Mason D, Strand V, Landewe R, Combe B (2008) Centrolizumab pegol with methotrexate significantly decreases signs and symptoms and progression of joint damage in patients with active rheumatoid arthritis: 1-year results from the rapid 1 trial. *Ann Rheum Dis* 67 (Suppl 11): 186

18 Smeets TJ, Kraan MC, van Loon ME, Tak PP (2003) Tumor necrosis factor alpha blockade reduces the synovial cell infiltrate early after initiation of treatment, but apparently not by induction of apoptosis in synovial tissue. *Arthritis Rheum* 48: 2155–62

19 Haringman JJ, Kraan MC, Smeets TJ, Zwinderman KH, Tak PP (2003) Chemokine blockade and chronic inflammatory disease: Proof of concept in patients with rheumatoid arthritis. *Ann Rheum Dis* 62: 715–21

20 Genovese MC, Cohen S, Moreland L, Lium D, Robbins S, Newmark R, Bekker P (2004) Combination therapy with etanercept and anakina in the treatment of patients with rheumatoid arthritis who have been treated unsuccessfully with methotrexate. *Arthritis Rheum* 50: 1412–19

21 Weinblatt M, Combe B, Covucci A, Aranda R, Becker JC, Keystone E (2006) Safety of the selective costimulatory modulator abatacept in rheumatoid arthritis patients receiving background biologic and monobiologic disease modifying antirheumatic drugs. A one year randomized placebo controlled study. *Arthritis Rheum* 54: 2807–16

22 Gardam MA, Keystone EC, Menzies R, Manners S, Skamene E, Long R, Vinh DC (2003) Anti-tumor necrosis factor agents and tuberculosis risk: Mechanisms of action and clinical management. *Lancet Infect Dis* 3: 148–155

23 Stone JH, Holbrook JT, Marriott MA, Tibbs AK, Sejismundo LP, Min YI, Specks U, Merkel PA, Spiera R, Davis JC (2006) Solid malignancies among patients in the Wegener's Granulomatosus Etanercept Trial. *Arthritis Rheum* 54: 1608–18

24 Bongartz I, Sutton AJ, Sweeting MJ, Buchan I, Matteson EL, Montori V (2006) Anti-TNF antibody therapy in rheumatoid arthritis and the risk of serious infections and malignancies systematic review and meta-analysis of rare harmful effects in randomized controlled trials. *JAMA* 295: 2275–85

25 Stuve O, Marra CM, Cravens PD, Singh MP, Hu W, Lovett-Racke A, Monson NL, Phillips JT, Tervaert JW, Nash RA et al (2007) Potential risk of progressive multifocal leukoencephalopathy with natalizumab therapy: Possible interventions. *Arch Neurol* 64: 169–76

26 Garcia-Suarez J, de Miguel D, Krsnik I, Banas H, Arribas I, Burgaleta C (2005) Changes in the natural history of progressive multifocal leukoencephalopahty in HIV negative lymphoproliferative disorders: Impact of novel therapies. *Am J Hematol* 80: 271–81

27 Strand V, Kimberly R, Isaacs JD (2007) Biologic therapies in rheumatology: Lessons learned, future directions. *Nat Rev Drug Discov* 6: 75–92

28 Maini R, St. Clair EW, Breedveld F, Furst D, Kalden J, Weisman M, Smolen J, Emery P, Harriman G, Feldmann M, Lipsky P (1999) Infliximab (chimeric anti-tumour necrosis factor alpha monoclonal antibody) *versus* placebo in rheumatoid arthritis patients receiving concomitant methotrexate: A randomized phase III trial. ATTRACT Study Group. *Lancet* 354: 1932–9

29 Keystone EC, Kavanaugh AF, Sharp JT, Tannenbaum H, Hua Y, Teoh LS, Fischkoff SA, Chartash EK (2004) Radiographic, clinical, and functional outcomes of treatment with adalimumab (a human anti-tumor necrosis factor monoclonal antibody) in patients with active rheumatoid arthritis receiving concomitant methotrexate therapy: A randomized, placebo-controlled, 52 week trial. *Arthritis Rheum* 50: 1400–11

30 Moreland LW, Schiff MH, Baumgartner SW, Tindall EA, Fleischmann RM, Bulpitt KJ, Weaver AL, Keystone EC, Furst DE, Mease PJ et al (1999) Etanercept therapy in rheumatoid arthritis. A randomized controlled trial. *Ann Intern Med* 130: 478–86

31 Choy EH, Isenberg DA, Garrood T, Farrow S, Ioannou Y, Bird H, Cheung N, Williams B, Hazleman B, Price R et al (2002) Therapeutic benefit of blocking interleukin-6 activity with an anti-interleukin-6 receptor monoclonal antibody in rheumatoid arthritis: A randomized, double-blind, placebo-controlled, dose-escalation trial. *Arthritis Rheum* 46: 3143–50

32 Cohen SB, Dore RK, Lane NE, Ory PA, Peterfy CG, Sharp JT, van der Heijde D, Zhou L, Tsuji W, Newmark R et al (2008) Denosumab treatment effects on structural damage, bone mineral density and bone turnover in rheumatoid arthritis: A twelve month, multicenter double-blind, placebo controlled, phase II clinical trial. *Arthritis Rheum* 58: 1299–309

33 Kremer JM, Genant HK, Moreland LW, Russell AS, Emery P, Abud-Mendoze C, Szechinski J, Li T, Ge Z, Becker JC, Westhovens R (2006) Effects of abatacept in patients with methotrexate-resistant active rheumatoid arthritis: A randomized trial. *Ann Intern Med* 144: 865–76

34 Cohen SB, Emery P, Greenwald MW, Dougados M, Furie RA, Genovese MC, Keystone EC, Loveless JE, Burmester GR, Cravets MW et al (2006) REFLEX Trial Group. Rituximab for rheumatoid arthritis refractory to anti-tumor necrosis factor therapy: Results of a multicenter, randomized, double-blind, placebo-controlled, phase III trial evaluating primary efficacy and safety at twenty-four weeks. *Arthritis Rheum* 54: 2793–806

35 Furst DE, Wallis R, Broder M, Beenhouwer DO (2006) Tumor necrosis factor antago-

nists: Different kinetics and/or mechanisms of action may explain differences in the risk for developing granulomatous infection. *Semin Arthritis Rheum* 36: 159–67

36 Schreiber S, Rutgeerts P, Fedorak RN, Khaliq-Kareemi M, Kamm MA, Boivin M, Bernstein CN, Staun M, Thomsen OO, Innes A et al (2005) A randomized, placebo-controlled trial of certolizumab pegol (BDP870) for treatment of Crohn's disease. *Gastroenterology* 129: 807–18

37 Hauser SL, Waubant E, Arnold DL, Vollmer T, Antel J, Fox RJ, Bar-Or A, Panzara M, Sarkar N, Agarwal S et al (2008) B-cell depletion with rituximab in relapsing-remitting multiple sclerosis. *N Engl J Med* 358: 676–88

38 Haraoui B, Cameron L, Ouellet M, White B (2006) Anti-infliximab antibodies in patients with rheumatoid arthritis who require higher doses of infliximab to achieve or maintain a clinical response. *J Rheumatol* 33: 31–6

39 Wolbink GJ, Vis M, Lems W, Voskuyl AE, de Groot E, Nurmohamed MT, Stapel S, Tak PP, Arden L, Dijkmans B (2006) Development of antiinfliximab antibodies and relationship to clinical response in patients with rheumatoid arthritis. *Arthritis Rheum* 54: 711–715

40 Bartelds GM, Wijbrandts CA, Nurmohamed MT, Stapel S, Lems WF, Aarden L, Dijkmans B, Tak PP, Wolbink GJ (2007) Clinical response to adalimumab: Relationship to anti-adalimumab antibodies and serum adalimumab concentrations in rheumatoid arthritis. *Ann Rheum Dis* 66: 921–926

41 Keystone EC (2003) Abandoned therapies and unpublished trials in rheumatoid arthritis. *Curr Opin Rheumatol* 15: 253–8

42 Epstein WV (1996) Expectation bias in rheumatoid arthritis clinical trials. The anti-CD4 monoclonal antibody experience. *Arthritis Rheum* 39: 1773–80

43 van der Heijde D, Klareskog L, Rodriguez-Valverde V, Codreanu C, Bolosiu H, Melo-Gomes J, Tornero-Molina J, Wajdula J, Pedersen R, Fatenejad S et al (2006) Comparison of etanercept and methotrexate, alone and combined in the treatment of rheumatoid arthritis: Two-year clinical and radiographic results from the TEMPO study, a double-blind, randomized trial. *Arthritis Rheum* 54: 1063–74

44 Breedveld FC, Weisman MH, Kavanaugh AF, Cohen SB, Pavelka K, van Vollenhoven R, Sharp J, Perez JL, Spencer-Green GT (2006) The PREMIER study: A multicenter randomized, double-blind clinical trial of combination therapy with adalimumab plus methotrexate *versus* methotrexate alone or adalimumab alone in patients with early, aggressive rheumatoid arthritis who had not had previous methotrexate therapy. *Arthritis Rheum* 54: 26–37

45 Smolen JS, Han C, Bala M, Maini RN, Kalden JR, van der Heijde D, Breedveld FC, Furst DE, Lipsky PE, ATTRACT Study Group (2005) Evidence of radiographic benefit of treatment with infliximab plus methotrexate in rheumatoid arthritis patients who had no clinical improvement: A detailed subanalysis of data from the anti-tumor necrosis factor trial in rheumatoid arthritis with concomitant therapy study. *Arthritis Rheum* 52: 1020–30

46 Goekoop-Ruit Erman YP, de Vries-Bouwstra JK, Allaart CF, van Zeben D, Kerstens

PJ, Hazes JM, Zwinderman AH, Peeters AJ, de Jonge-Bok JM, Mallée C et al (2007) Comparison of treatment strategies in early rheumatoid arthritis: A randomized trial. *Ann Intern Med* 146: 406–15

47 Klarenbeek NB, Guler-Yuksel M, van der Kooij, van der Heijde DM, Huizinga TW, Kerstens, PJ, Peeters AJ, Ronday HK, Westedt ML, Dijkmans BA et al (2008) Clinical outcomes of four different treatment strategies in patients with recent-onset rheumatoid arthritis: 5–year results of the best study. *Ann Rheum Dis* 66 (Suppl 11): 192

48 von Dongen H, van Aken, Lard LR, Visser K, Ronday HK, Hulsmans HM, Speyer I, Westedt ML, Peeters AJ, Allaart CF et al (2007) Efficacy of methotrexate treatment in patients with probable rheumatoid arthritis: A double-blind, randomized placebo-controlled trial. *Arthritis Rheum* 56: 1424–32

49 Keystone E, Chon Y, Eickenhorst T (2005) Comparison of efficacy responses to etanercept treatment in rheumatoid arthritis patients with moderate *versus* severe disease. *Ann Rheum Dis* 64: 180

Index

The PIR-Series
Progress in Inflammation Research

Homepage: www.birkhauser.ch

Up-to-date information on the latest developments in the pathology, mechanisms and therapy of inflammatory disease are provided in this monograph series. Areas covered include
vascular responses, skin inflammation, pain, neuroinflammation, arthritis cartilage and bone,
airways inflammation and asthma, allergy, cytokines and inflammatory mediators, cell signalling, and recent advances in drug therapy. Each volume is edited by acknowledged experts
providing succinct overviews on specific topics intended to inform and explain. The series is
of interest to academic and industrial biomedical researchers, drug development personnel
and rheumatologists, allergists, pathologists, dermatologists and other clinicians requiring
regular scientific updates.

Available volumes:
T Cells in Arthritis, P. Miossec, W. van den Berg, G. Firestein (Editors), 1998
Medicinal Fatty Acids, J. Kremer (Editor), 1998
Cytokines in Severe Sepsis and Septic Shock, H. Redl, G. Schlag (Editors), 1999
Cytokines and Pain, L. Watkins, S. Maier (Editors), 1999
Pain and Neurogenic Inflammation, S.D. Brain, P. Moore (Editors), 1999
Apoptosis and Inflammation, J.D. Winkler (Editor), 1999
Novel Inhibitors of Leukotrienes, G. Folco, B. Samuelsson, R.C. Murphy (Editors), 1999
Metalloproteinases as Targets for Anti-Inflammatory Drugs,
    K.M.K. Bottomley, D. Bradshaw, J.S. Nixon (Editors), 1999
Gene Therapy in Inflammatory Diseases, C.H. Evans, P. Robbins (Editors), 2000
Cellular Mechanisms in Airways Inflammation, C. Page, K. Banner, D. Spina (Editors), 2000
Inflammatory and Infectious Basis of Atherosclerosis, J.L. Mehta (Editor), 2001
Neuroinflammatory Mechanisms in Alzheimer's Disease. Basic and Clinical Research,
    J. Rogers (Editor), 2001
Inflammation and Stroke, G.Z. Feuerstein (Editor), 2001
NMDA Antagonists as Potential Analgesic Drugs,
    D.J.S. Sirinathsinghji, R.G. Hill (Editors), 2002
Mechanisms and Mediators of Neuropathic pain, A.B. Malmberg, S.R. Chaplan (Editors), 2002
Bone Morphogenetic Proteins. From Laboratory to Clinical Practice,
    S. Vukicevic, K.T. Sampath (Editors), 2002
The Hereditary Basis of Allergic Diseases, J. Holloway, S. Holgate (Editors), 2002
Inflammation and Cardiac Diseases, G.Z. Feuerstein, P. Libby, D.L. Mann (Editors), 2003
Mind over Matter – Regulation of Peripheral Inflammation by the CNS,
    M. Schäfer, C. Stein (Editors), 2003
Heat Shock Proteins and Inflammation, W. van Eden (Editor), 2003
Pharmacotherapy of Gastrointestinal Inflammation, A. Guglietta (Editor), 2004
Arachidonate Remodeling and Inflammation, A.N. Fonteh, R.L. Wykle (Editors), 2004
Recent Advances in Pathophysiology of COPD, P.J. Barnes, T.T. Hansel (Editors), 2004
Cytokines and Joint Injury, W.B. van den Berg, P. Miossec (Editors), 2004